LINEAR ALGEBRA

Steven Levandosky, Ph.D.
Stanford University

Pearson
Custom
Publishing

Cover Art: "Probability," by Angela Sciaraffa.

Printed in the United States of America

30 29 28 27 26 25 24 23 22 21 20

This manuscript was supplied camera-ready by the author.

Please visit our web site at www.pearsoncustom.com

ISBN 0–536–66747–0

BA 993786

PEARSON CUSTOM PUBLISHING
75 Arlington Street, Suite 300, Boston, MA 02116
A Pearson Education Company

Contents

1 Vectors in $\mathbf{R}^n$

The set of all real numbers is denoted by $\mathbf{R}$. The set of all ordered n-tuples of real numbers is denoted by $\mathbf{R}^n$. That is

$$\mathbf{R}^n = \{(x_1, x_2, \ldots, x_n) \mid x_i \in \mathbf{R} \text{ for } 1 \le i \le n\}.$$

We usually represent $\mathbf{R}$ by points on the number line, $\mathbf{R}^2$ by points in the plane, and $\mathbf{R}^3$ by points in space.

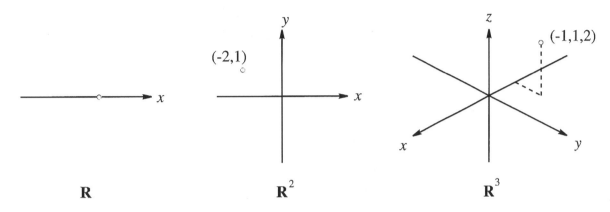

A **vector** in $\mathbf{R}^n$ is an ordered list of n real numbers

$$\mathbf{v} = \begin{bmatrix} v_1 \\ v_2 \\ \vdots \\ v_n \end{bmatrix},$$

where the numbers $v_1, v_2, \ldots, v_n$ are called the **components** of $\mathbf{v}$.

Example 1.1.

$$\mathbf{v} = \begin{bmatrix} 2 \\ -1 \end{bmatrix} \quad \text{and} \quad \mathbf{w} = \begin{bmatrix} 0 \\ 5 \end{bmatrix}$$

are vectors in $\mathbf{R}^2$ and

$$\mathbf{a} = \begin{bmatrix} \pi \\ 16 \\ -8 \\ 1 \end{bmatrix} \quad \text{and} \quad \mathbf{b} = \begin{bmatrix} 0 \\ 0 \\ 0 \\ 0 \end{bmatrix}$$

are vectors in $\mathbf{R}^4$. $\diamond$

The vector in $\mathbf{R}^n$ whose components are all zero is called the **zero vector** and is denoted by $\mathbf{0}$. There are two basic vector operations in $\mathbf{R}^n$, addition and scalar multiplication, both of which are performed *component-wise*. If $\mathbf{v}$ and $\mathbf{w}$ are vectors in $\mathbf{R}^n$ and c is a scalar (a real number), we define the sum of $\mathbf{v}$ and $\mathbf{w}$ by

$$\mathbf{v} + \mathbf{w} = \begin{bmatrix} v_1 \\ v_2 \\ \vdots \\ v_n \end{bmatrix} + \begin{bmatrix} w_1 \\ w_2 \\ \vdots \\ w_n \end{bmatrix} = \begin{bmatrix} v_1 + w_1 \\ v_2 + w_2 \\ \vdots \\ v_n + w_n \end{bmatrix},$$

and the scalar multiple c of $\mathbf{v}$ by

$$c\mathbf{v} = c \begin{bmatrix} v_1 \\ v_2 \\ \vdots \\ v_n \end{bmatrix} = \begin{bmatrix} cv_1 \\ cv_2 \\ \vdots \\ cv_n \end{bmatrix}.$$

Example 1.2.

$$2 \begin{bmatrix} 3 \\ 2 \\ -1 \end{bmatrix} + 3 \begin{bmatrix} 2 \\ 0 \\ 4 \end{bmatrix} = \begin{bmatrix} 6 \\ 4 \\ -2 \end{bmatrix} + \begin{bmatrix} 6 \\ 0 \\ 12 \end{bmatrix} = \begin{bmatrix} 12 \\ 4 \\ 10 \end{bmatrix}$$

$\Diamond$

We denote by $-\mathbf{v}$ the vector $(-1)\mathbf{v}$. The following properties of addition and scalar multiplication follow from the analogous properties of real numbers and are left as exercises.

Proposition 1.1. Let $\mathbf{u}$, $\mathbf{v}$ and $\mathbf{w}$ be vectors in $\mathbf{R}^n$ and let c and d be real numbers. Then

1. $\mathbf{u} + \mathbf{v} = \mathbf{v} + \mathbf{u}$
2. $\mathbf{u} + (\mathbf{v} + \mathbf{w}) = (\mathbf{u} + \mathbf{v}) + \mathbf{w}$
3. $\mathbf{v} + \mathbf{0} = \mathbf{v}$
4. $\mathbf{v} + (-\mathbf{v}) = \mathbf{0}$
5. $c(d\mathbf{v}) = (cd)\mathbf{v}$
6. $(c + d)\mathbf{v} = c\mathbf{v} + d\mathbf{v}$
7. $c(\mathbf{v} + \mathbf{w}) = c\mathbf{v} + c\mathbf{w}$
8. $1\mathbf{v} = \mathbf{v}$.

We represent vectors in $\mathbf{R}^2$ graphically by directed line segments. A nonzero vector

$$\mathbf{v} = \begin{bmatrix} v_1 \\ v_2 \end{bmatrix}$$

can be represented by a line segment starting at any point (x_1, x_2) in the plane and ending at the point $(x_1 + v_1, x_2 + v_2)$. The starting point is called the tail of $\mathbf{v}$ and the ending point is called the head of $\mathbf{v}$ and is denoted by an arrow.

Example 1.3. The vectors

$$\mathbf{u} = \begin{bmatrix} 1 \\ 2 \end{bmatrix} \qquad \mathbf{v} = \begin{bmatrix} 1 \\ -2 \end{bmatrix}$$

are shown below.

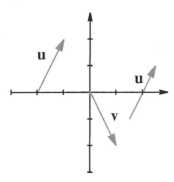

The vector **u** is shown in two different positions. A vector whose tail is at the origin is said to be in **standard position**. The vector **v** is shown in standard position. Notice that the point at the head of **v** is $(1, -2)$, whose coordinates are just the components of **v**. For this reason, it is common to identify a vector in standard position with the point at the head of the vector. ◇

Multiplication of a vector by a scalar $c \neq 1$ either lengthens or shortens the vector, and if the scalar is negative the direction of the vector is reversed.

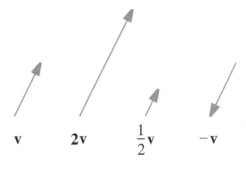

The sum of two vectors is represented geometrically by placing the vectors head to tail. The sum is the vector from the tail of the first vector to the head of the second vector.

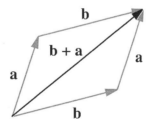

When the tails of **a** and **b** are at the same point, the difference vector **b** − **a** is represented by the vector whose tail is at the head of **a** and whose head is at the head of **b**. This is because **b** − **a** is the vector that, when added to **a**, equals **b**. Equivalently, **b** − **a** is the sum of **b** and −**a**.

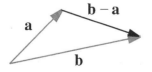

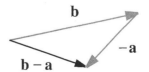

Vectors in $\mathbf{R}^3$ are also represented by directed line segments.

Example 1.4. The vector

$$\mathbf{v} = \begin{bmatrix} 2 \\ -1 \\ 3 \end{bmatrix}$$

is represented below in standard position.

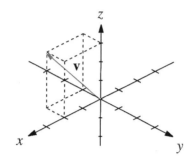

$\diamond$

Exercises

1.1. Compute the following.

(a) $2\begin{bmatrix} 1 \\ 2 \end{bmatrix} - \begin{bmatrix} 3 \\ 1 \end{bmatrix}$

(b) $2\left(\begin{bmatrix} 1 \\ 2 \end{bmatrix} - \begin{bmatrix} 3 \\ 1 \end{bmatrix}\right)$

1.2. Compute the following.

(a) $\begin{bmatrix} 1 \\ 3 \\ 2 \end{bmatrix} - 2\begin{bmatrix} -1 \\ 1 \\ 1 \end{bmatrix}$

(b) $3\begin{bmatrix} 4 \\ 0 \\ -1 \end{bmatrix} + 2\begin{bmatrix} 0 \\ 1 \\ 4 \end{bmatrix} - 3\begin{bmatrix} 1 \\ 1 \\ 0 \end{bmatrix}$

1.3. Let

$$\mathbf{v} = \begin{bmatrix} 3 \\ -2 \\ 3 \end{bmatrix} \qquad \mathbf{w} = \begin{bmatrix} 2 \\ 1 \\ -1 \end{bmatrix} \qquad \mathbf{x} = \begin{bmatrix} 1 \\ 4 \\ -5 \end{bmatrix}$$

Compute the following.

(a) $\mathbf{v} - 2\mathbf{w} + \mathbf{x}$

(b) $\mathbf{v} + \mathbf{w} + \mathbf{x}$

1.4. Let

$$\mathbf{a} = \begin{bmatrix} 1 \\ 2 \\ 3 \\ 4 \end{bmatrix} \qquad \mathbf{b} = \begin{bmatrix} 2 \\ 0 \\ 4 \\ 2 \end{bmatrix} \qquad \mathbf{c} = \begin{bmatrix} -1 \\ 3 \\ -2 \\ 4 \end{bmatrix}$$

Compute the following.

(a) $2\mathbf{a} + \mathbf{b} + \mathbf{c}$

(b) $2(\mathbf{a} + \mathbf{b}) - 3\mathbf{c}$

1.5. Sketch the following vectors in standard position.

(a) $\begin{bmatrix} 3 \\ 1 \end{bmatrix}$ (b) $\begin{bmatrix} -3 \\ -1 \end{bmatrix}$ (c) $\begin{bmatrix} -1 \\ 3 \end{bmatrix}$ (d) $\begin{bmatrix} 2 \\ 0 \end{bmatrix}$

1.6. Sketch the following vectors in standard position.

(a) $\begin{bmatrix} 2 \\ 1 \end{bmatrix}$ (b) $\begin{bmatrix} -2 \\ 1 \end{bmatrix}$ (c) $\begin{bmatrix} 0 \\ 0 \end{bmatrix}$ (d) $\begin{bmatrix} 0 \\ -2 \end{bmatrix}$

1.7. Let

$$\mathbf{a} = \begin{bmatrix} 2 \\ 1 \\ 1 \end{bmatrix} \qquad \mathbf{b} = \begin{bmatrix} 1 \\ -2 \\ 0 \end{bmatrix} \qquad \mathbf{c} = \begin{bmatrix} 4 \\ 2 \\ -2 \end{bmatrix} \qquad \mathbf{x} = \begin{bmatrix} 2 \\ 3 \end{bmatrix} \qquad \mathbf{y} = \begin{bmatrix} -3 \\ 2 \end{bmatrix}$$

Calculate the following.

(a) $\mathbf{a} + 2\mathbf{b} - \mathbf{c}$ (b) $\mathbf{a} + 2(\mathbf{b} - \mathbf{c})$ (c) $2\mathbf{x} - 3\mathbf{y}$ (d) $3\mathbf{x} + 2\mathbf{y}$

1.8. Consider the vectors shown below.

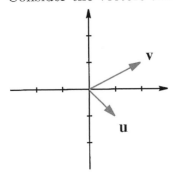

Sketch the following vectors in standard position.

(a) $-2\mathbf{u} + \mathbf{v}$

(b) $3\mathbf{u} + 2\mathbf{v}$

(c) $\mathbf{v} - \mathbf{u}$

(d) $\mathbf{u} - \mathbf{v}$.

1.9. Consider the vectors shown below.

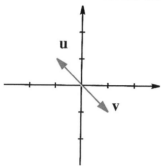

Sketch the following vectors in standard position.

    (a) $-2\mathbf{u} + \mathbf{v}$

    (b) $3\mathbf{u} + 2\mathbf{v}$

    (c) $\mathbf{v} - \mathbf{u}$

    (d) $\mathbf{u} - \mathbf{v}$.

1.10. Verify Properties 1-4 of Proposition 1.1.

1.11. Verify Properties 5-8 of Proposition 1.1.

2 Linear Combinations and Spans

Using the operations of addition and scalar multiplication, we can combine vectors to form new vectors. A **linear combination** of a set of vectors $\{\mathbf{v}_1, \mathbf{v}_2, \ldots, \mathbf{v}_k\}$ in $\mathbf{R}^n$ is a vector of the form

$$\mathbf{v} = c_1\mathbf{v}_1 + c_2\mathbf{v}_2 + \cdots + c_k\mathbf{v}_k$$

where the coefficients $c_1, \cdots, c_k$ are real numbers. Any or all of the coefficients could be zero.

Example 2.1. Let

$$\mathbf{v}_1 = \begin{bmatrix} 1 \\ 2 \\ 3 \end{bmatrix} \quad \text{and} \quad \mathbf{v}_2 = \begin{bmatrix} 2 \\ 0 \\ -1 \end{bmatrix}$$

Then

$$2\mathbf{v}_1 + 3\mathbf{v}_2 = \begin{bmatrix} 8 \\ 4 \\ 3 \end{bmatrix}$$

8

is a linear combination of $\mathbf{v}_1$ and $\mathbf{v}_2$. Both $\mathbf{v}_1$ and $\mathbf{v}_2$ are linear combinations of $\mathbf{v}_1$ and $\mathbf{v}_2$ since

$$\mathbf{v}_1 = 1\mathbf{v}_1 + 0\mathbf{v}_2$$
$$\mathbf{v}_2 = 0\mathbf{v}_1 + 1\mathbf{v}_2.$$

The zero vector in $\mathbf{R}^3$ is also a linear combination of $\mathbf{v}_1$ and $\mathbf{v}_2$ since $\mathbf{0} = 0\mathbf{v}_1 + 0\mathbf{v}_2$. $\diamond$

The **span** of a nonempty set of vectors is the set of all linear combinations of those vectors. That is,

$$\text{span}(\mathbf{v}_1, \mathbf{v}_2, \dots, \mathbf{v}_k) = \{c_1\mathbf{v}_1 + c_2\mathbf{v}_2 + \cdots + c_k\mathbf{v}_k \mid c_i \in \mathbf{R} \text{ for } 1 \leq i \leq k\}.$$

Example 2.2. Since $c\mathbf{0} = \mathbf{0}$ for any $c \in \mathbf{R}$, the span of the zero vector is itself. That is,

$$\text{span}(\mathbf{0}) = \{\mathbf{0}\}.$$

$\diamond$

Example 2.3. Let

$$\mathbf{v} = \begin{bmatrix} 1 \\ 2 \end{bmatrix}.$$

Then $\text{span}(\mathbf{v}) = \{c\mathbf{v} \mid c \in \mathbf{R}\}$ is the set of all scalar multiples of $\mathbf{v}$. When all of these vectors are placed in standard position, this set is represented by the line shown below.

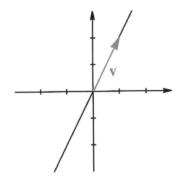

$\diamond$

The span of any nonzero vector is represented by a line passing through the origin, and conversely, any line which passes through the origin represents the span of some nonzero vector. By adding a fixed vector $\mathbf{x}_0$ to each vector in the span of some nonzero vector $\mathbf{v}$, we obtain the set $\{\mathbf{x}_0 + c\mathbf{v} \mid c \in \mathbf{R}\}$. When all of these vectors are placed in standard position, their heads represent points on the line which passes through $\mathbf{x}_0$ and is parallel to the line spanned by the vector $\mathbf{v}$.

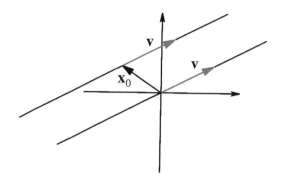

Thus we say that the set

$$L = \{\mathbf{x}_0 + t\mathbf{v} \mid t \in \mathbf{R}\}$$

is a **parametric** representation of the line in $\mathbf{R}^n$ which passes through $\mathbf{x}_0$ with direction $\mathbf{v}$. The variable t is called the **parameter**. Any given line may be parametrized in many ways. The point $\mathbf{x}_0$ can be any point on the line, and the direction vector $\mathbf{v}$ can be any nonzero vector whose head and tail are points in the line.

Example 2.4. Find a parametric representation of the line in $\mathbf{R}^3$ which passes through the points $(1, 1, 1)$ and $(2, -3, 6)$.
Solution. We may let

$$\mathbf{x}_0 = \begin{bmatrix} 1 \\ 1 \\ 1 \end{bmatrix}.$$

To obtain the direction of the line, we subtract the coordinates of the two points. So

$$\mathbf{v} = \begin{bmatrix} 2 \\ -3 \\ 6 \end{bmatrix} - \begin{bmatrix} 1 \\ 1 \\ 1 \end{bmatrix} = \begin{bmatrix} 1 \\ -4 \\ 5 \end{bmatrix}$$

and the line is given by

$$\left\{ \begin{bmatrix} 1 \\ 1 \\ 1 \end{bmatrix} + t \begin{bmatrix} 1 \\ -4 \\ 5 \end{bmatrix} \;\middle|\; t \in \mathbf{R} \right\}.$$

The components x, y and z of a point $\mathbf{x}$ on the line satisfy

$$x = 1 + t$$
$$y = 1 - 4t$$
$$z = 1 + 5t.$$

Each choice of t determines a point (x, y, z) on the line. $\diamond$

10

A **line segment** may be described by restricting the values of the parameter t in the parametric representation of the line containing the segment. The line segment which passes through the points $\mathbf{x}_0$ and $\mathbf{x}_1$ is contained in the line $\{\mathbf{x}_0 + t(\mathbf{x}_1 - \mathbf{x}_0) \mid t \in \mathbf{R}\}$. This parametrization yields $\mathbf{x}_0$ when $t = 0$, and $\mathbf{x}_1$ when $t = 1$. Thus the line segment between $\mathbf{x}_0$ and $\mathbf{x}_1$ is given by

$$S = \{\mathbf{x}_0 + t(\mathbf{x}_1 - \mathbf{x}_0) \mid 0 \le t \le 1\} = \{(1-t)\mathbf{x}_0 + t\mathbf{x}_1 \mid 0 \le t \le 1\}.$$

Next we consider the span of sets consisting of two vectors.

Example 2.5. Let

$$\mathbf{v}_1 = \begin{bmatrix} 1 \\ 2 \end{bmatrix} \quad \text{and} \quad \mathbf{v}_2 = \begin{bmatrix} 2 \\ 1 \end{bmatrix}.$$

Then $\text{span}(\mathbf{v}_1, \mathbf{v}_2) = \{c_1\mathbf{v}_1 + c_2\mathbf{v}_2 \mid c_1, c_2 \in \mathbf{R}\}$. It turns out that $\text{span}(\mathbf{v}_1, \mathbf{v}_2) = \mathbf{R}^2$. That is, *every* vector $\mathbf{x}$ in $\mathbf{R}^2$ can be expressed as a linear combination of $\mathbf{v}_1$ and $\mathbf{v}_2$. We will justify this assertion in two ways.

Geometrically. Consider some particular values of c_1 and c_2. First, fix $c_2 = 0$ and let c_1 vary over all of $\mathbf{R}$ to see that $\text{span}(\mathbf{v}_1, \mathbf{v}_2)$ contains the line spanned by $\mathbf{v}_1$. Next fix $c_2 = 1$ and let c_1 vary over $\mathbf{R}$ to see that $\text{span}(\mathbf{v}_1, \mathbf{v}_2)$ contains the line which passes through the head of $\mathbf{v}_2$ and is parallel to the line spanned by $\mathbf{v}_1$. By continuing to fix c_2 and let c_1 vary over $\mathbf{R}$ we see that $\text{span}(\mathbf{v}_1, \mathbf{v}_2)$ contains every line which passes through the line spanned by $\mathbf{v}_2$ and is parallel to the line spanned by $\mathbf{v}_1$.

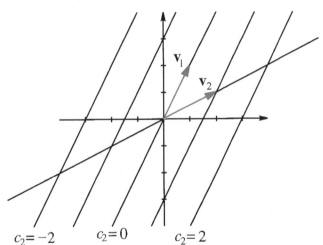

$c_2 = -2 \qquad c_2 = 0 \qquad c_2 = 2$

The union of all these lines is all of $\mathbf{R}^2$, so $\text{span}(\mathbf{v}_1, \mathbf{v}_2) = \mathbf{R}^2$.

Algebraically. Consider an arbitrary vector

$$\mathbf{x} = \begin{bmatrix} x_1 \\ x_2 \end{bmatrix}$$

in $\mathbf{R}^2$. We want to find scalars c_1 and c_2 such that $\mathbf{x} = c_1\mathbf{v}_1 + c_2\mathbf{v}_2$. That is,

$$\begin{bmatrix} x_1 \\ x_2 \end{bmatrix} = c_1 \begin{bmatrix} 1 \\ 2 \end{bmatrix} + c_2 \begin{bmatrix} 2 \\ 1 \end{bmatrix} = \begin{bmatrix} c_1 + 2c_2 \\ 2c_1 + c_2 \end{bmatrix}.$$

11

Equating the components of these vectors, we get

$$
\begin{aligned}
c_1 + 2c_2 &= x_1 \\
2c_1 + c_2 &= x_2,
\end{aligned}
$$

a system of two linear equations in the two unknowns c_1 and c_2. Such a system can be solved easily. If we subtract twice the first equation from the second equation, we get

$$
\begin{aligned}
c_1 + 2c_2 &= x_1 \\
-3c_2 &= x_2 - 2x_1.
\end{aligned}
$$

Solving the second equation for c_2 gives $c_2 = \frac{2}{3}x_1 - \frac{1}{3}x_2$. Substituting this into the first equation gives $c_1 = -\frac{1}{3}x_1 + \frac{2}{3}x_2$, and thus

$$
\mathbf{x} = \left(-\frac{1}{3}x_1 + \frac{2}{3}x_2\right)\mathbf{v}_1 + \left(\frac{2}{3}x_1 - \frac{1}{3}x_2\right)\mathbf{v}_2.
$$

This final expression gives an explicit formula for $\mathbf{x}$ as a linear combination of $\mathbf{v}_1$ and $\mathbf{v}_2$, so $\mathrm{span}(\mathbf{v}_1, \mathbf{v}_2) = \mathbf{R}^2$. $\diamond$

Example 2.6. Let

$$
\mathbf{v}_1 = \begin{bmatrix} 2 \\ 1 \end{bmatrix} \qquad \text{and} \qquad \mathbf{v}_2 = \begin{bmatrix} -4 \\ -2 \end{bmatrix},
$$

and notice that $\mathbf{v}_2 = -2\mathbf{v}_1$. Thus

$$
\mathrm{span}(\mathbf{v}_1, \mathbf{v}_2) = \{c_1\mathbf{v}_1 + c_2\mathbf{v}_2 \mid c_1, c_2 \in \mathbf{R}\} = \{(c_1 - 2c_2)\mathbf{v}_1 \mid c_1, c_2 \in \mathbf{R}\}.
$$

This set consists of all scalar multiples of $\mathbf{v}_1$ and is therefore a line, not all of $\mathbf{R}^2$.

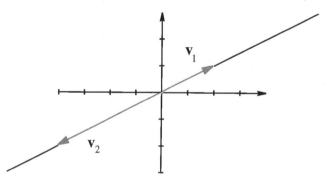

The fact that $\mathbf{v}_2$ is a scalar multiple of $\mathbf{v}_1$ means that $\mathbf{v}_2$ is in $\mathrm{span}(\mathbf{v}_1)$. In this sense, $\mathbf{v}_2$ is *redundant*, since adding multiples of $\mathbf{v}_2$ does not contribute anything new to the span. We could likewise regard $\mathbf{v}_1$ as redundant, since multiples of $\mathbf{v}_2$ alone are also sufficient to span the line. Thus, in this case $\mathrm{span}(\mathbf{v}_1, \mathbf{v}_2) = \mathrm{span}(\mathbf{v}_1) = \mathrm{span}(\mathbf{v}_2)$. $\diamond$

Two nonzero vectors are called **collinear** if one vector is a scalar multiple of the other. Notice that, if this is the case, then actually both vectors are scalar multiples of each other. For if $\mathbf{v} = c\mathbf{w}$, then the scalar c must be nonzero, so $\mathbf{w} = \frac{1}{c}\mathbf{v}$. It follows as in Example 2.6 that the span of any two collinear vectors is a line.

Example 2.7. Let

$$\mathbf{v}_1 = \begin{bmatrix} 1 \\ 2 \\ -1 \end{bmatrix} \qquad \text{and} \qquad \mathbf{v}_2 = \begin{bmatrix} 0 \\ 2 \\ 1 \end{bmatrix}.$$

Then

$$\mathrm{span}(\mathbf{v}_1, \mathbf{v}_2) = \{c_1\mathbf{v}_1 + c_2\mathbf{v}_2 \mid c_1, c_2 \in \mathbf{R}\}.$$

As in Example 2.5, $\mathrm{span}(\mathbf{v}_1, \mathbf{v}_2)$ contains every line which passes through $\mathrm{span}(\mathbf{v}_2)$ and is parallel to $\mathrm{span}(\mathbf{v}_1)$.

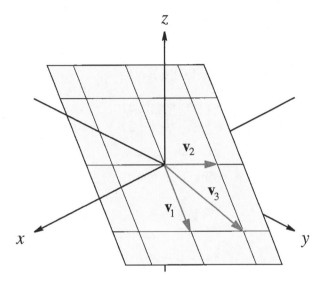

The union of all such lines forms a plane in $\mathbf{R}^3$ which passes through the origin and contains both $\mathbf{v}_1$ and $\mathbf{v}_2$. ◇

Generally, the span of any two nonzero, non-collinear vectors $\mathbf{v}_1$ and $\mathbf{v}_2$ in $\mathbf{R}^n$ ($n \geq 2$) is a plane which passes through the origin and contains the two spanning vectors. As we did with lines, we can represent arbitrary planes in $\mathbf{R}^n$ by adding a fixed vector $\mathbf{x}_0$ to vectors in this span. Thus we say that the set

$$P = \{\mathbf{x}_0 + s\mathbf{v}_1 + t\mathbf{v}_2 \mid s, t \in \mathbf{R}\}$$

is a parametric representation of the plane in $\mathbf{R}^n$ which passes through $\mathbf{x}_0$ and is parallel to $\mathrm{span}(\mathbf{v}_1, \mathbf{v}_2)$. The variables s and t are the parameters.

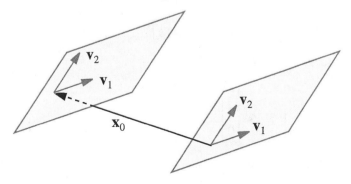

13

Example 2.8. Find a parametrization of the plane in $\mathbf{R}^3$ which contains the points $(1, 2, 3)$, $(2, -5, 4)$ and $(6, 4, 7)$.

Solution. We can first let

$$\mathbf{x}_0 = \begin{bmatrix} 1 \\ 2 \\ 3 \end{bmatrix}.$$

To obtain two direction vectors, we may use the vectors whose tails are at $(1, 2, 3)$ and whose heads are at $(2, -5, 4)$ and $(6, 4, 7)$ to get

$$\mathbf{v}_1 = \begin{bmatrix} 2 \\ -5 \\ 4 \end{bmatrix} - \begin{bmatrix} 1 \\ 2 \\ 3 \end{bmatrix} = \begin{bmatrix} 1 \\ -7 \\ 1 \end{bmatrix} \quad \text{and} \quad \mathbf{v}_2 = \begin{bmatrix} 6 \\ 4 \\ 7 \end{bmatrix} - \begin{bmatrix} 1 \\ 2 \\ 3 \end{bmatrix} = \begin{bmatrix} 5 \\ 2 \\ 4 \end{bmatrix}.$$

Thus the plane can be written parametrically as

$$\left\{ \begin{bmatrix} 1 \\ 2 \\ 3 \end{bmatrix} + s \begin{bmatrix} 1 \\ -7 \\ 1 \end{bmatrix} + t \begin{bmatrix} 5 \\ 2 \\ 4 \end{bmatrix} \;\middle|\; s, t \in \mathbf{R} \right\}$$

or in terms of the components of a point $\mathbf{x}$ in the plane,

$$x = 1 + s + 5t$$
$$y = 2 - 7s + 2t$$
$$z = 3 + s + 4t.$$

Each choice of the parameters s and t determines a point (x, y, z) on the plane. $\diamond$

Now we consider spans of three vectors.

Example 2.9. The vectors

$$\mathbf{e}_1 = \begin{bmatrix} 1 \\ 0 \\ 0 \end{bmatrix}, \qquad \mathbf{e}_2 = \begin{bmatrix} 0 \\ 1 \\ 0 \end{bmatrix}, \qquad \text{and} \qquad \mathbf{e}_3 = \begin{bmatrix} 0 \\ 0 \\ 1 \end{bmatrix}$$

span all of $\mathbf{R}^3$. To see this, let

$$\mathbf{v} = \begin{bmatrix} v_1 \\ v_2 \\ v_3 \end{bmatrix}$$

be an arbitrary vector in $\mathbf{R}^3$. Then

$$\mathbf{v} = v_1 \begin{bmatrix} 1 \\ 0 \\ 0 \end{bmatrix} + v_2 \begin{bmatrix} 0 \\ 1 \\ 0 \end{bmatrix} + v_3 \begin{bmatrix} 0 \\ 0 \\ 1 \end{bmatrix} = v_1 \mathbf{e}_1 + v_2 \mathbf{e}_2 + v_3 \mathbf{e}_3,$$

so $\mathbf{v}$ is a linear combination of $\mathbf{e}_1$, $\mathbf{e}_2$ and $\mathbf{e}_3$. Hence $\mathrm{span}(\mathbf{e}_1, \mathbf{e}_2, \mathbf{e}_3) = \mathbf{R}^3$. $\diamond$

Remark 2.1. The previous example can be extended naturally to $\mathbf{R}^n$ if we define $\mathbf{e}_j$ to be the vector in $\mathbf{R}^n$ whose components are all zero except for the jth component, which equals one. Then by the same reasoning as above it follows that $\mathbf{R}^n = \text{span}(\mathbf{e}_1, \mathbf{e}_2, \ldots, \mathbf{e}_n)$. In $\mathbf{R}^3$, the vectors $\mathbf{e}_1$, $\mathbf{e}_2$ and $\mathbf{e}_3$ are often denoted $\mathbf{i}$, $\mathbf{j}$ and $\mathbf{k}$, respectively, and vectors in $\mathbf{R}^3$ are commonly written as linear combinations of $\mathbf{i}$, $\mathbf{j}$ and $\mathbf{k}$. For instance, the vector

$$\begin{bmatrix} 2 \\ -4 \\ 7 \end{bmatrix}$$

is written $2\mathbf{i} - 4\mathbf{j} + 7\mathbf{k}$.

The next example shows that not every set of three vectors in $\mathbf{R}^3$ spans all of $\mathbf{R}^3$.

Example 2.10. Let

$$\mathbf{v}_1 = \begin{bmatrix} 1 \\ 2 \\ -1 \end{bmatrix}, \qquad \mathbf{v}_2 = \begin{bmatrix} 0 \\ 2 \\ 1 \end{bmatrix}, \qquad \text{and} \qquad \mathbf{v}_3 = \begin{bmatrix} 1 \\ 4 \\ 0 \end{bmatrix}.$$

Notice that $\mathbf{v}_3 = \mathbf{v}_1 + \mathbf{v}_2$. Therefore

$$\begin{aligned} \text{span}(\mathbf{v}_1, \mathbf{v}_2, \mathbf{v}_3) &= \{c_1\mathbf{v}_1 + c_2\mathbf{v}_2 + c_3\mathbf{v}_3 \mid c_1, c_2, c_3 \in \mathbf{R}\} \\ &= \{c_1\mathbf{v}_1 + c_2\mathbf{v}_2 + c_3(\mathbf{v}_1 + \mathbf{v}_2) \mid c_1, c_2, c_3 \in \mathbf{R}\} \\ &= \{(c_1 + c_3)\mathbf{v}_1 + (c_2 + c_3)\mathbf{v}_2 \mid c_1, c_2, c_3 \in \mathbf{R}\} \\ &= \text{span}(\mathbf{v}_1, \mathbf{v}_2). \end{aligned}$$

That is, since $\mathbf{v}_3$ is in the plane spanned by $\mathbf{v}_1$ and $\mathbf{v}_2$, adding multiples of $\mathbf{v}_3$ does not contribute anything new to the span, so $\text{span}(\mathbf{v}_1, \mathbf{v}_2, \mathbf{v}_3)$ is the same plane as in Example 2.7. For this reason we can regard $\mathbf{v}_3$ as redundant in the same sense that $\mathbf{v}_2$ was redundant in Example 2.6. Notice also that $\text{span}(\mathbf{v}_1, \mathbf{v}_2, \mathbf{v}_3) = \text{span}(\mathbf{v}_1, \mathbf{v}_3) = \text{span}(\mathbf{v}_2, \mathbf{v}_3)$. (Why?) $\diamond$

Exercises

2.1. Let

$$\mathbf{v} = \begin{bmatrix} 1 \\ 2 \end{bmatrix} \qquad \text{and} \qquad \mathbf{w} = \begin{bmatrix} 2 \\ 1 \end{bmatrix}.$$

(a) Write $\begin{bmatrix} 1 \\ 0 \end{bmatrix}$ as a linear combination of $\mathbf{v}$ and $\mathbf{w}$.

(b) Write $\begin{bmatrix} 0 \\ 1 \end{bmatrix}$ as a linear combination of $\mathbf{v}$ and $\mathbf{w}$.

(c) Write $\mathbf{x} = \begin{bmatrix} x_1 \\ x_2 \end{bmatrix}$ as a linear combination of $\begin{bmatrix} 1 \\ 0 \end{bmatrix}$ and $\begin{bmatrix} 0 \\ 1 \end{bmatrix}$.

(d) Write $\mathbf{x} = \begin{bmatrix} x_1 \\ x_2 \end{bmatrix}$ as a linear combination of $\mathbf{v}$ and $\mathbf{w}$.

2.2. Let

$$\mathbf{v} = \begin{bmatrix} 2 \\ 0 \\ 1 \end{bmatrix} \qquad \mathbf{w} = \begin{bmatrix} -1 \\ 2 \\ -2 \end{bmatrix}.$$

For each of the following vectors, either express it as a linear combination of $\mathbf{v}$ and $\mathbf{w}$, or explain why it is not a linear combination of $\mathbf{v}$ and $\mathbf{w}$.

(a) $\begin{bmatrix} 1 \\ 2 \\ 1 \end{bmatrix}$ (b) $\begin{bmatrix} 1 \\ 2 \\ -1 \end{bmatrix}$ (c) $\begin{bmatrix} 1 \\ 0 \\ 0 \end{bmatrix}$ (d) $\begin{bmatrix} 0 \\ 0 \\ 0 \end{bmatrix}$

For each set of vectors in Exercises 3 through 10 determine whether its span is a line, a plane or all of $\mathbf{R}^3$.

2.3. $\left\{ \begin{bmatrix} 2 \\ 1 \\ 1 \end{bmatrix}, \begin{bmatrix} 5 \\ 8 \\ 0 \end{bmatrix}, \begin{bmatrix} 1 \\ 6 \\ -2 \end{bmatrix} \right\}$

2.4. $\left\{ \begin{bmatrix} 2 \\ 1 \\ 1 \end{bmatrix}, \begin{bmatrix} 5 \\ 8 \\ 0 \end{bmatrix}, \begin{bmatrix} 2 \\ 6 \\ -2 \end{bmatrix} \right\}$

2.5. $\left\{ \begin{bmatrix} 1 \\ 0 \\ 1 \end{bmatrix}, \begin{bmatrix} 0 \\ 8 \\ 0 \end{bmatrix}, \begin{bmatrix} 2 \\ 2 \\ 2 \end{bmatrix} \right\}$

2.6. $\left\{ \begin{bmatrix} 1 \\ 2 \\ 3 \end{bmatrix}, \begin{bmatrix} 4 \\ 8 \\ 12 \end{bmatrix} \right\}$

2.7. $\left\{ \begin{bmatrix} 2 \\ 0 \\ 1 \end{bmatrix}, \begin{bmatrix} 6 \\ 3 \\ -3 \end{bmatrix}, \begin{bmatrix} 2 \\ 3 \\ -5 \end{bmatrix} \right\}$

2.8. $\left\{ \begin{bmatrix} 1 \\ 3 \\ -1 \end{bmatrix}, \begin{bmatrix} 2 \\ 6 \\ 2 \end{bmatrix} \right\}$

2.9. $\left\{ \begin{bmatrix} -2 \\ 3 \\ 1 \end{bmatrix}, \begin{bmatrix} 6 \\ -9 \\ -3 \end{bmatrix}, \begin{bmatrix} -4 \\ 6 \\ 2 \end{bmatrix} \right\}$

2.10. $\left\{ \begin{bmatrix} 1 \\ 2 \\ -3 \end{bmatrix}, \begin{bmatrix} 0 \\ 1 \\ 4 \end{bmatrix}, \begin{bmatrix} 0 \\ 0 \\ 1 \end{bmatrix} \right\}$

16

In Exercises 11 through 14 find a parametric representation of the line containing the given points.

2.11. $(3, 1)$ and $(2, -3)$

2.12. $(-4, 5)$ and $(0, 2)$

2.13. $(1, 2, 3)$ and $(-2, 1, 2)$

2.14. $(0, 4, 1)$ and $(2, -2, 4)$

In Exercises 15 and 16 find a parametric representation of the plane containing the given points.

2.15. $(1, 2, 3)$, $(2, 3, 4)$ and $(2, 1, 5)$

2.16. $(1, 1, 1)$, $(2, -3, 1)$ and $(4, 5, 2)$

2.17. Find a parametric form of the "hyperplane" in $\mathbf{R}^4$ which contains the points $(2, 3, 4, 5)$, $(1, 1, 1, 1)$, $(0, 1, 0, 1)$ and $(-1, -2, 3, 1)$. Hint: There are three direction vectors.

3 Linear Independence

We now give a precise meaning to the concept of redundancy of vectors illustrated in Examples 2.6 and 2.10. A set of vectors $\{\mathbf{v}_1, \mathbf{v}_2, \ldots, \mathbf{v}_k\}$ is called **linearly dependent** if at least one of the vectors is a linear combination of the others. Otherwise the set is called **linearly independent**. This definition applies to sets of two or more vectors. We say that a set consisting of a single vector $\mathbf{v}$ is linearly independent if $\mathbf{v} \neq \mathbf{0}$ and linearly dependent if $\mathbf{v} = \mathbf{0}$. For sets consisting of two vectors linear dependence simply means that one vector is a scalar multiple of the other. Any two nonzero, non-collinear vectors are linearly independent.

Example 3.1. In the figures below, $\mathbf{v}$ and $\mathbf{w}$ are linearly dependent, while $\mathbf{x}$ and $\mathbf{y}$ are linearly independent.

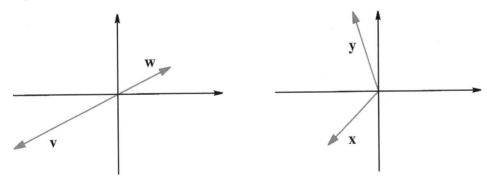

$\diamond$

Example 3.2. Let

$$\mathbf{u} = \begin{bmatrix} 1 \\ 2 \end{bmatrix}, \qquad \mathbf{v} = \begin{bmatrix} 3 \\ 2 \end{bmatrix}, \qquad \text{and} \qquad \mathbf{w} = \begin{bmatrix} 3 \\ -2 \end{bmatrix}.$$

The sets $\{\mathbf{u}, \mathbf{v}\}$, $\{\mathbf{u}, \mathbf{w}\}$ and $\{\mathbf{v}, \mathbf{w}\}$ are each linearly independent since no two of the vectors are collinear. The set $\{\mathbf{u}, \mathbf{v}, \mathbf{w}\}$, however, is linearly dependent since $\mathbf{w} = 2\mathbf{v} - 3\mathbf{u}$. We will see later that any set of three or more vectors in $\mathbf{R}^2$ is linearly dependent. $\diamond$

Note that the linear dependence in Example 3.2 is not entirely the fault of $\mathbf{w}$. We could also observe that $\mathbf{v} = \frac{1}{2}\mathbf{w} + \frac{3}{2}\mathbf{u}$, so $\mathbf{v}$ is dependent on $\mathbf{u}$ and $\mathbf{w}$, or $\mathbf{u} = \frac{2}{3}\mathbf{v} - \frac{1}{3}\mathbf{w}$ so $\mathbf{u}$ is dependent on $\mathbf{v}$ and $\mathbf{w}$. It is the fact that there are linear relations among the three vectors that makes $\{\mathbf{u}, \mathbf{v}, \mathbf{w}\}$ a linearly dependent set. An equivalent formulation of linear dependence is the following symmetric condition.

> **Proposition 3.1.** A set of vectors $\{\mathbf{v}_1, \mathbf{v}_2, \dots, \mathbf{v}_k\}$ is linearly dependent if and only if
>
> $$c_1\mathbf{v}_1 + c_2\mathbf{v}_2 + \cdots + c_k\mathbf{v}_k = \mathbf{0} \tag{3.1}$$
>
> for some scalars $c_1, c_2, \dots c_k$, not all of which are zero.

In other words, a set $\{\mathbf{v}_1, \mathbf{v}_2, \dots, \mathbf{v}_k\}$ is linearly independent if the *only* linear combination of $\mathbf{v}_1$ through $\mathbf{v}_k$ which equals $\mathbf{0}$ is the combination with all the coefficients equal to zero, called the **trivial** combination.

Proof. First suppose a set of vectors is linearly dependent. Then one of the vectors (say $\mathbf{v}_1$) equals a linear combination of the others. So

$$\mathbf{v}_1 = d_2\mathbf{v}_2 + \cdots + d_k\mathbf{v}_k.$$

Subtracting $\mathbf{v}_1$ from both sides results in

$$(-1)\mathbf{v}_1 + d_2\mathbf{v}_2 + \cdots + d_k\mathbf{v}_k = \mathbf{0}$$

which is an equation of the form (3.1) with not all the coefficients equal to zero. On the other hand, suppose (3.1) holds with at least one nonzero coefficient. Suppose $c_1 \neq 0$. Then solving for $\mathbf{v}_1$ gives

$$\mathbf{v}_1 = -\frac{c_2}{c_1}\mathbf{v}_2 - \cdots - \frac{c_k}{c_1}\mathbf{v}_k$$

so $\mathbf{v}_1$ is a linear combination of the other vectors, and the set $\{\mathbf{v}_1, \mathbf{v}_2, \ldots, \mathbf{v}_k\}$ is linearly dependent. □

Example 3.3. Let

$$\mathbf{v}_1 = \begin{bmatrix} 1 \\ 2 \\ 2 \end{bmatrix}, \qquad \mathbf{v}_2 = \begin{bmatrix} 1 \\ 3 \\ 1 \end{bmatrix}, \qquad \text{and} \qquad \mathbf{v}_3 = \begin{bmatrix} -1 \\ -5 \\ 1 \end{bmatrix}.$$

Is $\{\mathbf{v}_1, \mathbf{v}_2, \mathbf{v}_3\}$ a linearly independent set?

Solution. To answer this question we need to determine if there exist real numbers c_1, c_2 and c_3 (not all zero) such that $c_1\mathbf{v}_1 + c_2\mathbf{v}_2 + c_3\mathbf{v}_3 = \mathbf{0}$. Equating the components of this vector equation we get the equations

$$\begin{array}{rrrrrl} c_1 & + & c_2 & - & c_3 & = 0 \\ 2c_1 & + & 3c_2 & - & 5c_3 & = 0 \\ 2c_1 & + & c_2 & + & c_3 & = 0. \end{array}$$

This is called a **system of linear equations**. To solve this system we use the method of **elimination**. The idea is to use one of the equations to eliminate one of the variables from the other equations. If we subtract twice the first equation from both the second and third equations, we get the new (but equivalent) system

$$\begin{array}{rrrrrl} c_1 & + & c_2 & - & c_3 & = 0 \\ & & c_2 & - & 3c_3 & = 0 \\ & & -c_2 & + & 3c_3 & = 0. \end{array}$$

Adding the second equation to the third gives

$$\begin{array}{rrrrrl} c_1 & + & c_2 & - & c_3 & = 0 \\ & & c_2 & - & 3c_3 & = 0 \\ & & & & 0 & = 0. \end{array}$$

Thus the third equation reduces to the trivially true statement $0 = 0$, so we can ignore it. To find a solution of the first two equations, notice that, given any choice of c_3, we can solve for c_1 and c_2. Thus we are free to choose c_3 however we desire. A simple choice is $c_3 = 1$ which leads to $c_2 = 3$ from the second equation, and then $c_1 = -2$ from the first equation. It is easy to check that these values satisfy all three of the original equations. Thus

$$-2\mathbf{v}_1 + 3\mathbf{v}_2 + \mathbf{v}_3 = \mathbf{0}$$

which means that $\{\mathbf{v}_1, \mathbf{v}_2, \mathbf{v}_3\}$ is a linearly dependent set. ◇

Example 3.4. The vectors $\mathbf{e}_1, \ldots, \mathbf{e}_n$ defined in Remark 2.1 are linearly independent. To see this, suppose

$$c_1\mathbf{e}_1 + c_2\mathbf{e}_2 + \cdots + c_n\mathbf{e}_n = \mathbf{0}.$$

Writing out the components, this equation becomes

$$\begin{bmatrix} c_1 \\ c_2 \\ \vdots \\ c_n \end{bmatrix} = \begin{bmatrix} 0 \\ 0 \\ \vdots \\ 0 \end{bmatrix}$$

so all of the coefficients must equal zero, and thus the vectors are linearly independent. ◇

Exercises

For each set in Exercises 1 through 6 show that the set is linearly independent or express one of the vectors in the set as a linear combination of the others.

3.1. $\left\{ \begin{bmatrix} 2 \\ 1 \end{bmatrix}, \begin{bmatrix} 3 \\ 2 \end{bmatrix} \right\}$

3.2. $\left\{ \begin{bmatrix} 2 \\ 1 \end{bmatrix}, \begin{bmatrix} 3 \\ 2 \end{bmatrix}, \begin{bmatrix} 1 \\ 2 \end{bmatrix} \right\}$

3.3. $\left\{ \begin{bmatrix} 2 \\ 1 \\ 1 \end{bmatrix}, \begin{bmatrix} 1 \\ 1 \\ 0 \end{bmatrix}, \begin{bmatrix} 1 \\ 0 \\ 1 \end{bmatrix}, \begin{bmatrix} 0 \\ 0 \\ 1 \end{bmatrix} \right\}$

3.4. $\left\{ \begin{bmatrix} 1 \\ -1 \\ 2 \end{bmatrix}, \begin{bmatrix} 2 \\ 1 \\ 3 \end{bmatrix}, \begin{bmatrix} -1 \\ 0 \\ 2 \end{bmatrix} \right\}$

3.5. $\left\{ \begin{bmatrix} 2 \\ 0 \end{bmatrix}, \begin{bmatrix} 0 \\ 3 \end{bmatrix}, \begin{bmatrix} 5 \\ 7 \end{bmatrix} \right\}$

3.6. $\left\{ \begin{bmatrix} 1 \\ -1 \\ 3 \end{bmatrix}, \begin{bmatrix} 2 \\ 1 \\ 3 \end{bmatrix}, \begin{bmatrix} 3 \\ 3 \\ 3 \end{bmatrix} \right\}$

3.7. Let $\{\mathbf{u}, \mathbf{v}, \mathbf{w}\}$ be a linearly independent set. Show that $\{\mathbf{u} + \mathbf{v}, \mathbf{u} + \mathbf{w}, \mathbf{v} + \mathbf{w}\}$ is a linearly independent set.

3.8. Let $\{\mathbf{u}, \mathbf{v}, \mathbf{w}\}$ be a linearly independent set. Is $\{\mathbf{u} - \mathbf{v}, \mathbf{v} - \mathbf{w}, \mathbf{u} - \mathbf{w}\}$ a linearly independent set? Show that it is or show why it is not.

3.9. Let $\{\mathbf{u}, \mathbf{v}, \mathbf{w}\}$ be a linearly independent set. Is $\{2\mathbf{u} + \mathbf{v}, \mathbf{u} + \mathbf{v} + \mathbf{w}, 2\mathbf{v} + 3\mathbf{w}\}$ a linearly independent set? Show that it is or show why it is not.

True/False. For Exercises 10 through 13, determine whether the given statement is true or false. If true, show why. If false, provide a counterexample.

3.10. If $S = \{\mathbf{v}_1, \dots, \mathbf{v}_k\}$ is a set of linearly independent vectors in $\mathbf{R}^n$, then any subset of S must be linearly independent.

3.11. If $S = \{\mathbf{v}_1, \ldots, \mathbf{v}_k\}$ is a set of linearly dependent vectors in $\mathbf{R}^n$, then any subset of S must be linearly dependent.

3.12. If $\text{span}(\mathbf{v}_1, \mathbf{v}_2, \mathbf{v}_3) = \mathbf{R}^3$, then $\{\mathbf{v}_1, \mathbf{v}_2, \mathbf{v}_3\}$ must be a linearly independent set.

3.13. If $S = \{\mathbf{v}_1, \mathbf{v}_2, \mathbf{v}_3\}$ is a linearly dependent set, then every vector in S can be written as a linear combination of the other 2 vectors.

4 Dot Products and Cross Products

The **dot product** of two vectors $\mathbf{v}$ and $\mathbf{w}$ in $\mathbf{R}^n$ is defined to be

$$\mathbf{v} \cdot \mathbf{w} = \begin{bmatrix} v_1 \\ v_2 \\ \vdots \\ v_n \end{bmatrix} \cdot \begin{bmatrix} w_1 \\ w_2 \\ \vdots \\ w_n \end{bmatrix} = v_1 w_1 + v_2 w_2 + \cdots + v_n w_n.$$

Notice that the dot product of two vectors is a scalar (real number), not a vector.

Example 4.1. Let

$$\mathbf{v} = \begin{bmatrix} 2 \\ 1 \end{bmatrix} \quad \mathbf{w} = \begin{bmatrix} -1 \\ 3 \end{bmatrix} \quad \mathbf{a} = \begin{bmatrix} 4 \\ 2 \\ 1 \end{bmatrix} \quad \mathbf{b} = \begin{bmatrix} 0 \\ -1 \\ 2 \end{bmatrix}.$$

Then $\mathbf{v} \cdot \mathbf{w} = 2(-1) + 1(3) = 1$ and $\mathbf{a} \cdot \mathbf{b} = 4(0) + 2(-1) + 1(2) = 0$. $\diamond$

The following properties of the dot product are easy to check.

Proposition 4.1. For any vectors $\mathbf{v}$, $\mathbf{w}$ and $\mathbf{x}$ in $\mathbf{R}^n$ and any scalar c,

1. $\mathbf{v} \cdot \mathbf{w} = \mathbf{w} \cdot \mathbf{v}$

2. $(\mathbf{v} + \mathbf{w}) \cdot \mathbf{x} = \mathbf{v} \cdot \mathbf{x} + \mathbf{w} \cdot \mathbf{x}$

3. $(c\mathbf{v}) \cdot \mathbf{w} = c(\mathbf{v} \cdot \mathbf{w}).$

There is a very close relationship between dot products and the notions of lengths and angles in $\mathbf{R}^n$. We define the **length** of a vector $\mathbf{v}$ in $\mathbf{R}^n$ to be

$$\|\mathbf{v}\| = \sqrt{v_1^2 + v_2^2 + \cdots + v_n^2}.$$

The expression under the square root is the dot product of $\mathbf{v}$ with itself, so

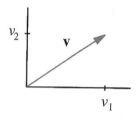

$$\mathbf{v} \cdot \mathbf{v} = \|\mathbf{v}\|^2.$$

The length of $\mathbf{v}$ is also commonly called the **magnitude** or **norm** of $\mathbf{v}$. The zero vector has length zero, and by the Pythagorean Theorem, the length of any nonzero vector $\mathbf{v} = (v_1, v_2)$ in $\mathbf{R}^2$ is just the length of the line segment which represents $\mathbf{v}$.

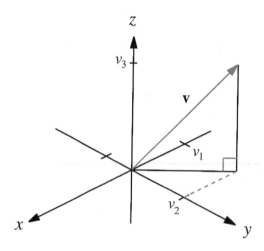

The same holds for vectors in $\mathbf{R}^3$. The segment which represents the nonzero vector $\mathbf{v} = (v_1, v_2, v_3)$ shown below is the hypotenuse of a right triangle.

The leg in the xy-plane has length $\sqrt{v_1^2 + v_2^2}$, so the length of the hypotenuse is

$$\sqrt{\left(\sqrt{v_1^2 + v_2^2}\right)^2 + v_3^2} = \sqrt{v_1^2 + v_2^2 + v_3^2},$$

the length of $\mathbf{v}$. The following properties are left as exercises.

Proposition 4.2.

1. $\|\mathbf{v}\| \geq 0$ for any $\mathbf{v}$ in $\mathbf{R}^n$.

2. $\|\mathbf{v}\| = 0$ if and only if $\mathbf{v} = \mathbf{0}$.

3. $\|c\mathbf{v}\| = |c|\|\mathbf{v}\|$ for any $\mathbf{v}$ in $\mathbf{R}^n$ and any scalar c.

A vector with length 1 is called a **unit vector**. Given any nonzero vector $\mathbf{v}$, if we apply the third property above with $c = 1/\|\mathbf{v}\|$, it follows that the vector

$$\mathbf{u} = \frac{\mathbf{v}}{\|\mathbf{v}\|}$$

is a unit vector in the same direction as $\mathbf{v}$.

The following result, known as the Cauchy-Schwarz inequality, provides an upper bound on the size of $\mathbf{v} \cdot \mathbf{w}$ in terms of the lengths of $\mathbf{v}$ and $\mathbf{w}$.

Proposition 4.3. (Cauchy-Schwarz Inequality) For any nonzero vectors $\mathbf{v}, \mathbf{w}$ in $\mathbf{R}^n$,

$$|\mathbf{v} \cdot \mathbf{w}| \leq \|\mathbf{v}\|\|\mathbf{w}\|.$$

Furthermore, we have the equality $|\mathbf{v} \cdot \mathbf{w}| = \|\mathbf{v}\|\|\mathbf{w}\|$ if and only if $\mathbf{v} = c\mathbf{w}$ for some nonzero scalar c.

Proof. Let $\mathbf{v}$ and $\mathbf{w}$ be any nonzero vectors in $\mathbf{R}^n$ and define $p(t) = \|t\mathbf{w} - \mathbf{v}\|^2$ for any real number t. Using the properties listed above, we have

$$\begin{aligned} p(t) &= \|t\mathbf{w} - \mathbf{v}\|^2 \\ &= (t\mathbf{w} - \mathbf{v}) \cdot (t\mathbf{w} - \mathbf{v}) \\ &= t\mathbf{w} \cdot (t\mathbf{w} - \mathbf{v}) - \mathbf{v} \cdot (t\mathbf{w} - \mathbf{v}) \\ &= t^2 \mathbf{w} \cdot \mathbf{w} - 2t\mathbf{v} \cdot \mathbf{w} + \mathbf{v} \cdot \mathbf{v} \\ &= \|\mathbf{w}\|^2 t^2 - 2(\mathbf{v} \cdot \mathbf{w})t + \|\mathbf{v}\|^2 \\ &= at^2 - bt + c \end{aligned}$$

where $a = \|\mathbf{w}\|^2$, $b = 2(\mathbf{v} \cdot \mathbf{w})$ and $c = \|\mathbf{v}\|^2$. Since $\|\mathbf{v} - t\mathbf{w}\|^2$ is nonnegative, it follows that $p(t)$ is nonnegative for all t. In particular, $p(b/2a) \geq 0$ (the quantity $b/2a$ is defined since a is positive). This gives

$$a\left(\frac{b}{2a}\right)^2 - b\left(\frac{b}{2a}\right) + c = -\frac{b^2}{4a} + c \geq 0.$$

Since a is positive, this implies $b^2 \leq 4ac$, so

$$4(\mathbf{v} \cdot \mathbf{w})^2 \leq 4\|\mathbf{v}\|^2\|\mathbf{w}\|^2.$$

Dividing by 4 and taking the square root of both sides proves the inequality.

Next suppose $\mathbf{v} = c\mathbf{w}$ for some nonzero scalar c. Then

$$|\mathbf{v} \cdot \mathbf{w}| = |c\mathbf{w} \cdot \mathbf{w}| = |c|\|\mathbf{w}\|^2 = |c|\|\mathbf{w}\|\|\mathbf{w}\| = \|c\mathbf{w}\|\|\mathbf{w}\| = \|\mathbf{v}\|\|\mathbf{w}\|.$$

Conversely, suppose that we have the equality $|\mathbf{v} \cdot \mathbf{w}| = \|\mathbf{v}\|\|\mathbf{w}\|$. Then $b^2 = 4ac$, so

$$p(b/2a) = -\frac{b^2}{4a} + c = 0$$

23

which implies that

$$\left\| \mathbf{v} - \frac{b}{2a}\mathbf{w} \right\| = 0.$$

Since the only vector with norm zero is the zero vector, it follows that

$$\mathbf{v} = \frac{b}{2a}\mathbf{w}.$$

Since $\mathbf{v}$ and $\mathbf{w}$ are nonzero, $\mathbf{v}$ is a nonzero scalar multiple of $\mathbf{w}$. $\qquad\square$

An immediate consequence of the Cauchy-Schwarz inequality is the triangle inequality.

Proposition 4.4. (Triangle Inequality) Let $\mathbf{v}$ and $\mathbf{w}$ be nonzero vectors in $\mathbf{R}^n$. Then

$$\|\mathbf{v} + \mathbf{w}\| \le \|\mathbf{v}\| + \|\mathbf{w}\|,$$

and we have the equality $\|\mathbf{v} + \mathbf{w}\| = \|\mathbf{v}\| + \|\mathbf{w}\|$ if and only if $\mathbf{v} = c\mathbf{w}$ for some positive scalar c.

Geometrically this is equivalent to the statement that the length of any side of a triangle is no larger than the sum of the lengths of the other two sides.

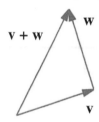

Proof. Using the properties of dot products, we have

$$
\begin{aligned}
\|\mathbf{v} + \mathbf{w}\|^2 &= (\mathbf{v} + \mathbf{w}) \cdot (\mathbf{v} + \mathbf{w}) \\
&= \mathbf{v} \cdot \mathbf{v} + 2\mathbf{v} \cdot \mathbf{w} + \mathbf{w} \cdot \mathbf{w} \\
&\le \mathbf{v} \cdot \mathbf{v} + 2\|\mathbf{v}\|\|\mathbf{w}\| + \mathbf{w} \cdot \mathbf{w} \\
&= \|\mathbf{v}\|^2 + 2\|\mathbf{v}\|\|\mathbf{w}\| + \|\mathbf{w}\|^2 \\
&= (\|\mathbf{v}\| + \|\mathbf{w}\|)^2
\end{aligned}
$$

where the Cauchy-Schwarz inequality was applied in the third line. Taking the square root of both sides proves the inequality.

Observe that we have equality if and only if $\mathbf{v} \cdot \mathbf{w} = \|\mathbf{v}\|\|\mathbf{w}\|$. By the second statement in the Cauchy-Schwarz inequality, this means that $\mathbf{v} = c\mathbf{w}$ for some $c \ne 0$. Then

$$c\|\mathbf{w}\|^2 = c\mathbf{w} \cdot \mathbf{w} = \mathbf{v} \cdot \mathbf{w} = \|\mathbf{v}\|\|\mathbf{w}\|$$

so c is positive. On the other hand, if $\mathbf{v} = c\mathbf{w}$ for some positive scalar c, then

$$\mathbf{v} \cdot \mathbf{w} = c\mathbf{w} \cdot \mathbf{w} = c\|\mathbf{w}\|^2 = c\|\mathbf{w}\|\|\mathbf{w}\| = \|c\mathbf{w}\|\|\mathbf{w}\| = \|\mathbf{v}\|\|\mathbf{w}\|$$

and therefore we have equality. $\qquad\square$

Using the triangle inequality it is possible to make sense of the angle between two vectors in $\mathbf{R}^n$. Let $\mathbf{v}$ and $\mathbf{w}$ be nonzero vectors in $\mathbf{R}^n$ and assume that $\mathbf{v}$ is not a scalar multiple of $\mathbf{w}$. We can associate to this pair of vectors a triangle in the plane with side lengths $A = \|\mathbf{v}\|$, $B = \|\mathbf{w}\|$ and $C = \|\mathbf{v} - \mathbf{w}\|$.

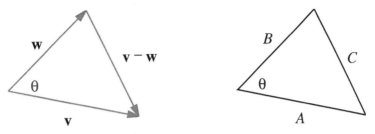

To see that there exists such a triangle we apply the triangle inequality three times. First

$$\|\mathbf{v} - \mathbf{w}\| = \|\mathbf{v} + (-\mathbf{w})\| < \|\mathbf{v}\| + \|-\mathbf{w}\| = \|\mathbf{v}\| + \|\mathbf{w}\|$$

so $C < A + B$. Next

$$\|\mathbf{v}\| = \|\mathbf{w} + (\mathbf{v} - \mathbf{w})\| < \|\mathbf{w}\| + \|\mathbf{v} - \mathbf{w}\|$$

so $A < B + C$. Finally

$$\|\mathbf{w}\| = \|\mathbf{v} + (\mathbf{w} - \mathbf{v})\| < \|\mathbf{v}\| + \|\mathbf{w} - \mathbf{v}\| = \|\mathbf{v}\| + \|\mathbf{v} - \mathbf{w}\|$$

so $B < A + C$. By the SSS Theorem of plane geometry, all such triangles are congruent. We can therefore *define* the angle θ between $\mathbf{v}$ and $\mathbf{w}$ to be the corresponding angle in any plane triangle with sides of length $\|\mathbf{v}\|$, $\|\mathbf{w}\|$ and $\|\mathbf{v} - \mathbf{w}\|$. If $\mathbf{v} = c\mathbf{w}$ then we define $\theta = 0$ if c is positive and $\theta = \pi$ if c is negative. Dot products provide us with an easy way to determine the angle between two vectors.

Proposition 4.5. Let $\mathbf{v}, \mathbf{w}$ be nonzero vectors in $\mathbf{R}^n$. Then

$$\mathbf{v} \cdot \mathbf{w} = \|\mathbf{v}\|\|\mathbf{w}\| \cos\theta$$

where θ is the angle between $\mathbf{v}$ and $\mathbf{w}$.

Proof. Consider the plane triangle with side lengths $A = \|\mathbf{v}\|$, $B = \|\mathbf{w}\|$ and $C = \|\mathbf{v} - \mathbf{w}\|$. The Law of Cosines states that

$$C^2 = A^2 + B^2 - 2AB \cos\theta.$$

That is,

$$\|\mathbf{v} - \mathbf{w}\|^2 = \|\mathbf{v}\|^2 + \|\mathbf{w}\|^2 - 2\|\mathbf{v}\|\|\mathbf{w}\| \cos\theta. \tag{4.1}$$

Using the properties of dot products, the left side of equation (4.1) may be rewritten

$$\|\mathbf{v} - \mathbf{w}\|^2 = (\mathbf{v} - \mathbf{w}) \cdot (\mathbf{v} - \mathbf{w})$$
$$= \mathbf{v} \cdot (\mathbf{v} - \mathbf{w}) - \mathbf{w} \cdot (\mathbf{v} - \mathbf{w})$$
$$= \mathbf{v} \cdot \mathbf{v} - \mathbf{v} \cdot \mathbf{w} - \mathbf{w} \cdot \mathbf{v} + \mathbf{w} \cdot \mathbf{w}$$
$$= \|\mathbf{v}\|^2 + \|\mathbf{w}\|^2 - 2(\mathbf{v} \cdot \mathbf{w}).$$

Comparing this with the right side of equation (4.1) proves the identity. □

We say two nonzero vectors are **perpendicular** if the angle between them is a right angle. In this case $\cos\theta = 0$. Thus Proposition 4.5 implies the following test for perpendicularity.

Proposition 4.6. Two nonzero vectors $\mathbf{v}$ and $\mathbf{w}$ in $\mathbf{R}^n$ are perpendicular if and only if $\mathbf{v} \cdot \mathbf{w} = 0$.

We say that two vectors $\mathbf{v}$ and $\mathbf{w}$ are **orthogonal** if $\mathbf{v} \cdot \mathbf{w} = 0$. The distinction between the terms perpendicular and orthogonal arises only with regard to the zero vector. Perpendicularity does not make sense when one of the vectors is the zero vector, since angles are only defined between nonzero vectors. Thus we never say that the zero vector is perpendicular to another vector. However, the dot product of the zero vector in $\mathbf{R}^n$ with any other vector in $\mathbf{R}^n$ makes sense, and is always zero, so the zero vector is orthogonal to every vector in $\mathbf{R}^n$. The following is the vector form of the Pythagorean Theorem. Its proof is left as an exercise.

Proposition 4.7. Suppose $\mathbf{v}$ and $\mathbf{w}$ are orthogonal. Then

$$\|\mathbf{v} + \mathbf{w}\|^2 = \|\mathbf{v}\|^2 + \|\mathbf{w}\|^2.$$

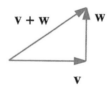

Dot products may be used to describe planes in $\mathbf{R}^3$. Given a plane, any vector perpendicular to the plane is called a **normal vector** to the plane. For instance, the vector $\mathbf{k}$ is a normal vector to the xy-plane.

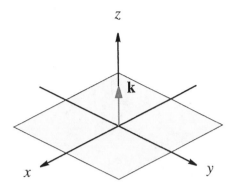

Suppose we know that a plane has a normal vector $\mathbf{n}$ and contains the point $\mathbf{x}_0$. If $\mathbf{x}$ is any other point in the plane then the difference $\mathbf{x} - \mathbf{x}_0$ is a vector parallel to the plane, and therefore perpendicular to $\mathbf{n}$.

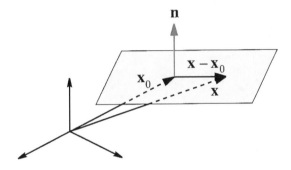

Thus by Proposition 4.6, $\mathbf{n} \cdot (\mathbf{x} - \mathbf{x}_0) = 0$. Using the notation

$$\mathbf{x} = \begin{bmatrix} x \\ y \\ z \end{bmatrix} \qquad \mathbf{x}_0 = \begin{bmatrix} x_0 \\ y_0 \\ z_0 \end{bmatrix} \qquad \mathbf{n} = \begin{bmatrix} n_1 \\ n_2 \\ n_3 \end{bmatrix}$$

the relation $\mathbf{n} \cdot (\mathbf{x} - \mathbf{x}_0) = 0$ takes the form

$$n_1(x - x_0) + n_2(y - y_0) + n_3(z - z_0) = 0.$$

This is the equation of the plane passing through $\mathbf{x}_0$ with normal vector $\mathbf{n}$.

Example 4.2. If a plane passes through the point $(1, 2, 3)$ and has normal vector

$$\mathbf{n} = \begin{bmatrix} 1 \\ 3 \\ -2 \end{bmatrix},$$

then its equation is $1(x - 1) + 3(y - 2) - 2(z - 3) = 0$, or equivalently $x + 3y - 2z = 1$. $\diamond$

Often the difficult part is finding a normal vector.

Example 4.3. Find the equation of the plane containing the points $(1, 2, 3)$, $(2, -5, 4)$ and $(6, 4, 7)$.

Solution. We may choose

$$\mathbf{x}_0 = \begin{bmatrix} 1 \\ 2 \\ 3 \end{bmatrix}.$$

To obtain two vectors parallel to the plane, we may use the vectors between $(1, 2, 3)$ and the other two points. Computing the difference vectors, we find

$$\mathbf{v} = \begin{bmatrix} 1 \\ -7 \\ 1 \end{bmatrix} \qquad \mathbf{w} = \begin{bmatrix} 5 \\ 2 \\ 4 \end{bmatrix}$$

are parallel to the plane. Thus any normal vector $\mathbf{n} = \begin{bmatrix} n_1 \\ n_2 \\ n_3 \end{bmatrix}$ must be perpendicular to both $\mathbf{v}$ and $\mathbf{w}$. So

$$\mathbf{v} \cdot \mathbf{n} = 0 \quad \text{and} \quad \mathbf{w} \cdot \mathbf{n} = 0.$$

This leads to the system of equations

$$
\begin{aligned}
1n_1 \;-\; 7n_2 \;+\; 1n_3 &= 0 \\
5n_1 \;+\; 2n_2 \;+\; 4n_3 &= 0.
\end{aligned}
$$

Subtracting 5 times the first equation from the second equation we get

$$
\begin{aligned}
n_1 \;-\; 7n_2 \;+\; 1n_3 &= 0 \\
37n_2 \;-\; n_3 &= 0.
\end{aligned}
$$

Any choice of n_3 will allow us to solve the second equation for n_2 and then the first equation for n_1. In view of the second equation it makes sense to choose $n_3 = 37$, so that from the second equation we get $n_2 = 1$, and then from the third equation $n_1 = -30$. So

$$\mathbf{n} = \begin{bmatrix} -30 \\ 1 \\ 37 \end{bmatrix}$$

is a normal vector and thus the equation of the plane is

$$-30(x - 1) + 1(y - 2) + 37(z - 3) = 0.$$

$\diamond$

A quick way to find a normal vector is to use cross products. The **cross product** of two vectors $\mathbf{v}$ and $\mathbf{w}$ in $\mathbf{R}^3$ is the vector

$$\mathbf{v} \times \mathbf{w} = \begin{bmatrix} v_2 w_3 - v_3 w_2 \\ v_3 w_1 - v_1 w_3 \\ v_1 w_2 - v_2 w_1 \end{bmatrix}.$$

The following properties are left as exercises.

Proposition 4.8. For any $\mathbf{u}$, $\mathbf{v}$ and $\mathbf{w}$ in $\mathbf{R}^3$

1. $\mathbf{v} \times \mathbf{w} = -\mathbf{w} \times \mathbf{v}$,

2. $(\mathbf{u} + \mathbf{v}) \times \mathbf{w} = \mathbf{u} \times \mathbf{w} + \mathbf{v} \times \mathbf{w}$,

3. $(c\mathbf{v}) \times \mathbf{w} = c(\mathbf{v} \times \mathbf{w})$,

4. $\mathbf{v} \cdot (\mathbf{v} \times \mathbf{w}) = 0$,

5. $\mathbf{w} \cdot (\mathbf{v} \times \mathbf{w}) = 0$.

The final two properties imply that the cross product of $\mathbf{v}$ and $\mathbf{w}$ is orthogonal to both $\mathbf{v}$ and $\mathbf{w}$. The direction of $\mathbf{v} \times \mathbf{w}$ is determined by the **right hand rule**. With the right hand positioned so that the fingers point in the direction of $\mathbf{v}$ and the palm faces toward $\mathbf{w}$, the thumb points in the direction of $\mathbf{v} \times \mathbf{w}$.

Example 4.4. Consider the plane from Example 4.3, with direction vectors

$$\mathbf{v} = \begin{bmatrix} 1 \\ -7 \\ 1 \end{bmatrix} \quad \text{and} \quad \mathbf{w} = \begin{bmatrix} 5 \\ 2 \\ 4 \end{bmatrix}.$$

Their cross product

$$\mathbf{v} \times \mathbf{w} = \begin{bmatrix} (-7)(4) - (1)(2) \\ (1)(5) - (1)(4) \\ (1)(2) - (-7)(5) \end{bmatrix} = \begin{bmatrix} -30 \\ 1 \\ 37 \end{bmatrix}$$

is precisely the normal vector we found earlier. $\diamond$

The length of $\mathbf{v} \times \mathbf{w}$ also depends on the lengths of $\mathbf{v}$ and $\mathbf{w}$ and the angle between $\mathbf{v}$ and $\mathbf{w}$.

Proposition 4.9. Let $\mathbf{v}$ and $\mathbf{w}$ be nonzero vectors in $\mathbf{R}^3$. Then

$$\|\mathbf{v} \times \mathbf{w}\| = \|\mathbf{v}\|\|\mathbf{w}\| \sin \theta$$

where θ is the angle between $\mathbf{v}$ and $\mathbf{w}$.

Proof. First,

$$\|\mathbf{v} \times \mathbf{w}\|^2 = (v_2 w_3 - v_3 w_2)^2 + (v_3 w_1 - v_1 w_3)^2 + (v_1 w_2 - v_2 w_1)^2$$
$$= v_1^2(w_2^2 + w_3^2) + v_2^2(w_1^2 + w_3^2) + v_3^2(w_1^2 + w_2^2)$$
$$- 2(v_1 w_1 v_2 w_2 + v_1 w_1 v_3 w_3 + v_2 v_2 v_3 w_3).$$

Next, squaring both sides of the relation $\mathbf{v} \cdot \mathbf{w} = \|\mathbf{v}\|\|\mathbf{w}\| \cos\theta$ gives

$$\|\mathbf{v}\|^2\|\mathbf{w}\|^2 \cos^2\theta = (v_1 w_1 + v_2 w_2 + v_3 w_3)^2$$
$$= v_1^2 w_1^2 + v_2^2 w_2^2 + v_3^2 w_3^2 + 2(v_1 w_1 v_2 w_2 + v_1 w_1 v_3 w_3 + v_2 v_2 v_3 w_3).$$

Adding these equations gives

$$\|\mathbf{v} \times \mathbf{w}\|^2 + \|\mathbf{v}\|^2\|\mathbf{w}\|^2 \cos^2\theta = (v_1^2 + v_2^2 + v_3^2)(w_1^2 + w_2^2 + w_3^2) = \|\mathbf{v}\|^2\|\mathbf{w}\|^2$$

so

$$\|\mathbf{v} \times \mathbf{w}\|^2 = \|\mathbf{v}\|^2\|\mathbf{w}\|^2(1 - \cos^2\theta) = \|\mathbf{v}\|^2\|\mathbf{w}\|^2 \sin^2\theta.$$

Taking the square root of both sides proves the identity. $\qquad\square$

This identity can be interpreted as a statement about area. Consider the parallelogram formed by $\mathbf{v}$ and $\mathbf{w}$.

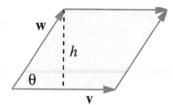

The width of the parallelogram is $\|\mathbf{v}\|$ and the height is $h = \|\mathbf{w}\| \sin\theta$, so the area is $\|\mathbf{v}\|\|\mathbf{w}\| \sin\theta$. Thus by Proposition 4.9, we have the following result.

Proposition 4.10. The area of the parallelogram formed by $\mathbf{v}$ and $\mathbf{w}$ is $\|\mathbf{v} \times \mathbf{w}\|$.

Example 4.5. Find the area of the triangle in $\mathbf{R}^3$ with vertices $(1,1,1)$, $(2,-1,3)$ and $(0,4,1)$.

Solution. The sides adjacent to the vertex $(1,1,1)$ are described by the vectors

$$\mathbf{v} = \begin{bmatrix} 1 \\ -2 \\ 2 \end{bmatrix} \qquad \text{and} \qquad \mathbf{w} = \begin{bmatrix} -1 \\ 3 \\ 0 \end{bmatrix}.$$

The triangle is therefore half of the parallelogram generated by $\mathbf{v}$ and $\mathbf{w}$. Since

$$\mathbf{v} \times \mathbf{w} = \begin{bmatrix} -6 \\ -2 \\ 1 \end{bmatrix}$$

the area of the triangle is $\frac{1}{2}\|\mathbf{v} \times \mathbf{w}\| = \frac{1}{2}\sqrt{41}$. $\qquad\diamond$

Exercises

4.1. Let

$$\mathbf{a} = \begin{bmatrix} 2 \\ 1 \\ 1 \end{bmatrix} \qquad \mathbf{b} = \begin{bmatrix} 1 \\ -2 \\ 0 \end{bmatrix} \qquad \mathbf{c} = \begin{bmatrix} 4 \\ 2 \\ -2 \end{bmatrix} \qquad \mathbf{x} = \begin{bmatrix} 2 \\ 3 \end{bmatrix} \qquad \mathbf{y} = \begin{bmatrix} -3 \\ 2 \end{bmatrix}$$

Calculate the following.
  (a) $\mathbf{a} \cdot \mathbf{b}$ (b) $\mathbf{a} \cdot (\mathbf{b} + \mathbf{c})$ (c) $\mathbf{x} \cdot \mathbf{y}$ (d) $\|\mathbf{x} - \mathbf{y}\|$

4.2. Let

$$\mathbf{u} = \begin{bmatrix} 1 \\ 2 \\ -1 \\ -2 \end{bmatrix} \qquad \mathbf{v} = \begin{bmatrix} -2 \\ 0 \\ 4 \\ 3 \end{bmatrix} \qquad \mathbf{w} = \begin{bmatrix} 1 \\ 6 \\ 1 \\ -3 \end{bmatrix}$$

Calculate the following.
  (a) $\mathbf{u} \cdot (\mathbf{w} - \mathbf{v})$ (b) $\mathbf{u} \cdot \mathbf{w} - \mathbf{u} \cdot \mathbf{v}$ (c) $\|\mathbf{u} + \mathbf{v}\|$
  (d) $\|\mathbf{u}\| + \|\mathbf{v}\|$ (e) $\mathbf{v} \cdot \mathbf{w}$ (f) $(2\mathbf{v}) \cdot \mathbf{w}$ (g) $\mathbf{v} \cdot (2\mathbf{w})$

4.3. Find the length of each of the following vectors.

  (a) $\begin{bmatrix} 1 \\ -1 \end{bmatrix}$ (b) $\begin{bmatrix} 4 \\ 3 \end{bmatrix}$ (c) $\begin{bmatrix} 2 \\ 1 \\ -2 \end{bmatrix}$ (d) $\begin{bmatrix} 1 \\ 1 \\ 1 \\ 1 \end{bmatrix}$

4.4. Find a vector in $\mathbf{R}^2$ with length 3 which points in the same direction as $\begin{bmatrix} 1 \\ -3 \end{bmatrix}$.

4.5. Find a vector in $\mathbf{R}^3$ which points in the opposite direction as $\begin{bmatrix} 2 \\ -7 \\ 1 \end{bmatrix}$, but is three times as long.

4.6. A force $\mathbf{F}_1$ of magnitude 3N is applied in the direction $\begin{bmatrix} -1 \\ 2 \end{bmatrix}$. In what direction should a force $\mathbf{F}_2$ of magnitude 2N be applied in order for the net force to have first component zero?

4.7. Suppose an airplane is pointing due west and flying with an airspeed of 500 mph, and that there is a 20 mph wind out of the southwest. Find the velocity vector of the plane relative to the ground. What is the speed of the plane relative to the ground?

4.8. Simplify the following expressions.

  (a) $(\mathbf{a} + \mathbf{b}) \cdot (\mathbf{c} + \mathbf{d})$

(b) $(\mathbf{a} + \mathbf{b}) \cdot (\mathbf{a} - \mathbf{b})$

4.9. (a) Simplify $\|\mathbf{x} + \mathbf{y}\|^2 - \|\mathbf{x}\|^2 - \|\mathbf{y}\|^2$ using dot products.

(b) Suppose $\|\mathbf{w}\| = \|\mathbf{z}\|$. Use dot products to show that $\mathbf{w} + \mathbf{z}$ and $\mathbf{w} - \mathbf{z}$ are orthogonal.

4.10. Suppose $\mathbf{u}$, $\mathbf{v}$ and $\mathbf{w}$ are nonzero vectors in $\mathbf{R}^n$ which are mutually orthogonal. That is, $\mathbf{u} \cdot \mathbf{v} = 0$, $\mathbf{u} \cdot \mathbf{w} = 0$ and $\mathbf{v} \cdot \mathbf{w} = 0$. Show that $\{\mathbf{u}, \mathbf{v}, \mathbf{w}\}$ is a linearly independent set.

4.11. Suppose that

$$\begin{array}{lll} \mathbf{u} \cdot \mathbf{x} = 1 & \mathbf{u} \cdot \mathbf{y} = 2 & \mathbf{u} \cdot \mathbf{z} = 0 \\ \mathbf{v} \cdot \mathbf{x} = -1 & \mathbf{v} \cdot \mathbf{y} = 2 & \mathbf{v} \cdot \mathbf{z} = 3 \\ \mathbf{w} \cdot \mathbf{x} = 1 & \mathbf{w} \cdot \mathbf{y} = -2 & \mathbf{w} \cdot \mathbf{z} = -1 \end{array}$$

Compute $(\mathbf{u} - \mathbf{v} + 2\mathbf{w}) \cdot (\mathbf{x} + \mathbf{y} - \mathbf{z})$.

4.12. Suppose that

$$\begin{array}{lll} \mathbf{x} \cdot \mathbf{y} = 3 & \mathbf{x} \cdot \mathbf{z} = -1 & \mathbf{y} \cdot \mathbf{z} = 2 \\ \|\mathbf{x}\| = 2 & \|\mathbf{y}\| = 3 & \|\mathbf{z}\| = 4 \end{array}$$

Compute the following.

(a) $(\mathbf{x} + \mathbf{y}) \cdot (\mathbf{x} + \mathbf{z})$

(b) $\|\mathbf{x} + \mathbf{y} + \mathbf{z}\|$

4.13. Let

$$\mathbf{u} = \begin{bmatrix} 1 \\ -2 \\ 0 \end{bmatrix} \qquad \mathbf{v} = \begin{bmatrix} 1 \\ 2 \\ 3 \end{bmatrix} \qquad \mathbf{w} = \begin{bmatrix} 2 \\ 1 \\ -1 \end{bmatrix}$$

Compute

(a) $\mathbf{u} \times \mathbf{v}$

(b) $\mathbf{v} \times \mathbf{u}$

(c) $\mathbf{v} \times \mathbf{w}$

(d) $(\mathbf{u} \times \mathbf{v}) \times \mathbf{w}$

(e) $\mathbf{u} \times (\mathbf{u} \times \mathbf{w})$

(f) $\mathbf{u} \times \mathbf{u}$

In Exercises 14 and 15, find the angle between the two vectors.

4.14. $\mathbf{u} = (1, 2)$ and $\mathbf{v} = (3, 4)$.

4.15. $\mathbf{u} = (-2, 1)$ and $\mathbf{v} = (4, 2)$.

4.16. Find a vector perpendicular to $\mathbf{v} = (1, 3)$.

4.17. Find two linearly independent vectors perpendicular to $\mathbf{v} = (1, 2, 3)$.

In Exercises 18 and 19 find (a) a normal vector to the plane containing the given points, and (b) an equation for the plane in terms of x, y and z.

4.18. $(1, 2, 3)$, $(2, 3, 4)$ and $(2, 1, 5)$.

4.19. $(1, 1, 1)$, $(2, -3, 1)$ and $(4, 5, 2)$.

4.20. (a) Show that $\mathbf{v} \cdot \mathbf{w} = \mathbf{w} \cdot \mathbf{v}$ for any vectors $\mathbf{v}$ and $\mathbf{w}$ in $\mathbf{R}^n$.

   (b) Show that $(\mathbf{v} + \mathbf{w}) \cdot \mathbf{x} = \mathbf{v} \cdot \mathbf{x} + \mathbf{w} \cdot \mathbf{x}$ for any vectors $\mathbf{v}$, $\mathbf{w}$ and $\mathbf{x}$ in $\mathbf{R}^n$.

   (c) Show that $(c\mathbf{v}) \cdot \mathbf{w} = c(\mathbf{v} \cdot \mathbf{w})$ for any vectors $\mathbf{v}$ and $\mathbf{w}$ in $\mathbf{R}^n$ and any scalar c in $\mathbf{R}$.

4.21. (a) Show that $\|\mathbf{v}\| \geq 0$ for any $\mathbf{v}$ in $\mathbf{R}^n$.

   (b) Show that $\|\mathbf{v}\| = 0$ if and only if $\mathbf{v} = \mathbf{0}$.

   (c) Show that $\|c\mathbf{v}\| = |c| \|\mathbf{v}\|$ for any $\mathbf{v}$ in $\mathbf{R}^n$ and any scalar c in $\mathbf{R}$.

4.22. Prove the following properties of cross products.

   (a) $\mathbf{v} \times \mathbf{w} = -\mathbf{w} \times \mathbf{v}$

   (b) $(\mathbf{u} + \mathbf{v}) \times \mathbf{w} = \mathbf{u} \times \mathbf{w} + \mathbf{v} \times \mathbf{w}$

   (c) $(c\mathbf{v}) \times \mathbf{w} = c(\mathbf{v} \times \mathbf{w})$

   (d) $\mathbf{v} \cdot (\mathbf{v} \times \mathbf{w}) = 0$

   (e) $\mathbf{w} \cdot (\mathbf{v} \times \mathbf{w}) = 0$

4.23. Find the area of the parallelogram in $\mathbf{R}^2$ with vertices $(1, 1)$, $(3, 2)$, $(4, 3)$ and $(2, 2)$.

4.24. Find the area of the triangle in $\mathbf{R}^3$ with vertices $(-2, 1, 2)$, $(3, 4, 1)$ and $(2, -1, 0)$.

4.25. Prove the Pythagorean Theorem for vectors. That is, show that if $\mathbf{v}$ and $\mathbf{w}$ are orthogonal then $\|v + w\|^2 = \|v\|^2 + \|w\|^2$.

4.26. Let P be the **parallelopiped** generated by vectors $\mathbf{u}$, $\mathbf{v}$ and $\mathbf{w}$. Show that the volume of P is $|\mathbf{u} \cdot (\mathbf{v} \times \mathbf{w})|$. *Hint: The volume of a region with uniform cross sectional area is its base area times its height.*

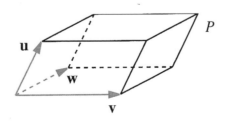

5 Systems of Linear Equations

We have encountered systems of linear equations when trying to answer the following questions.

- Is a given vector in the span of some collection of vectors?

- Given a collection of vectors, are they linearly independent?

- What vectors are perpendicular to a given collection of vectors?

The general form of a linear system of m equations in n unknowns is

$$
\begin{array}{ccccccccc}
a_{11}x_1 & + & a_{12}x_2 & + & \cdots & + & a_{1n}x_n & = & b_1 \\
a_{21}x_1 & + & a_{22}x_2 & + & \cdots & + & a_{2n}x_n & = & b_2 \\
\vdots & & \vdots & & & & \vdots & & \vdots \\
a_{m1}x_1 & + & a_{m2}x_2 & + & \cdots & + & a_{mn}x_n & = & b_m
\end{array}
$$

where the coefficients a_{ij} and the right hand sides b_i are given and we wish to solve for the unknowns x_1 through x_n.

The standard technique for solving systems of linear equations is called **Gaussian elimination**. The idea is to use one equation to eliminate one of the variables from the other equations. We illustrate the process in the following examples.

Example 5.1. Find all solutions of the system

$$
\begin{array}{ccccccc}
x_1 & - & x_2 & + & x_3 & = & -2 \\
3x_1 & - & x_2 & - & x_3 & = & 8 \\
2x_1 & + & x_2 & - & 2x_3 & = & 11.
\end{array}
$$

Solution. First, we use the top equation to eliminate x_1 from the other equations. Subtracting 3 times the first equation from the second equation and subtracting 2 times the first equation from the third equation gives

$$
\begin{array}{ccccccc}
x_1 & - & x_2 & + & x_3 & = & -2 \\
 & & 2x_2 & - & 4x_3 & = & 14 \\
 & & 3x_2 & - & 4x_3 & = & 15.
\end{array}
$$

Next use the new middle equation to eliminate x_2 from the other equations. First divide by 2 to make the calculations simpler.

$$
\begin{array}{ccccccc}
x_1 & - & x_2 & + & x_3 & = & -2 \\
 & & x_2 & - & 2x_3 & = & 7 \\
 & & 3x_2 & - & 4x_3 & = & 15.
\end{array}
$$

Adding the second equation to the first and subtracting 3 times the second equation from the third gives

$$
\begin{array}{ccccccc}
x_1 & & & - & x_3 & = & 5 \\
 & & x_2 & - & 2x_3 & = & 7 \\
 & & & & 2x_3 & = & -6.
\end{array}
$$

34

Dividing the new third equation by 2 gives

$$\begin{aligned} x_1 \quad - \quad x_3 &= 5 \\ x_2 - 2x_3 &= 7 \\ x_3 &= -3. \end{aligned}$$

Finally, adding the third equation to the first and adding 2 times the third equation to the second gives

$$\begin{aligned} x_1 \quad\quad\quad &= 2 \\ x_2 \quad\quad &= 1 \\ x_3 &= -3. \end{aligned}$$

Thus the solution is

$$\begin{bmatrix} x_1 \\ x_2 \\ x_3 \end{bmatrix} = \begin{bmatrix} 2 \\ 1 \\ -3 \end{bmatrix}.$$

Geometrically this solution is the point of intersection of the three planes described by the original equations.

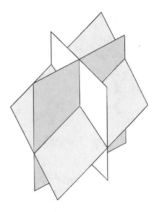

$\diamond$

Example 5.2. Find all solutions of the system

$$\begin{aligned} x_1 - x_2 + x_3 &= 2 \\ 4x_1 - 3x_2 + x_3 &= 3. \end{aligned}$$

Solution. Proceeding as in Example 5.1, we use the first equation to eliminate x_1 from the second equation to get

$$\begin{aligned} x_1 - x_2 + x_3 &= 2 \\ x_2 - 3x_3 &= -5. \end{aligned}$$

Adding the second equation to the first then yields

$$\begin{aligned} x_1 \quad\quad - 2x_3 &= -3 \\ x_2 - 3x_3 &= -5. \end{aligned}$$

Since there is no third equation with which to eliminate x_3 from these two equations, the process terminates here. For each choice of x_3 these equations can be solved separately for x_1 and x_2. Thus, by letting x_3 vary over $\mathbf{R}$ we obtain an infinite set of solutions. Solving for x_1 and x_2 we get $x_1 = -3 + 2x_3$ and $x_2 = -5 + 3x_3$. So the solutions take the form

$$\begin{bmatrix} x_1 \\ x_2 \\ x_3 \end{bmatrix} = \begin{bmatrix} -3 + 2x_3 \\ -5 + 3x_3 \\ x_3 \end{bmatrix} = \begin{bmatrix} -3 \\ -5 \\ 0 \end{bmatrix} + x_3 \begin{bmatrix} 2 \\ 3 \\ 1 \end{bmatrix}.$$

This is precisely the parametric representation of a line in $\mathbf{R}^3$, the intersection of the planes described by the two equations.

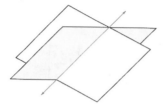

The set of solutions is infinite since the two equations do not provide enough information to uniquely determine a solution. $\diamond$

Example 5.3. Find all solutions of the system

$$\begin{array}{rrrrrrr}
x_1 & + & x_2 & + & 2x_3 & = & 1 \\
x_1 & + & 2x_2 & + & x_3 & = & 2 \\
2x_1 & + & 4x_2 & + & x_3 & = & 3 \\
3x_1 & - & 2x_2 & + & 4x_3 & = & 4.
\end{array}$$

Using the first equation to eliminate x_1 from the other three gives

$$\begin{array}{rrrrrrr}
x_1 & + & x_2 & + & 2x_3 & = & 1 \\
& & x_2 & - & x_3 & = & 1 \\
& & 2x_2 & - & 3x_3 & = & 1 \\
& & -5x_2 & - & 2x_3 & = & 1.
\end{array}$$

Using the second equation to eliminate x_2 from the other three gives

$$\begin{array}{rrrrrr}
x_1 & & + & 3x_3 & = & 0 \\
& x_2 & - & x_3 & = & 1 \\
& & & -x_3 & = & -1 \\
& & & -7x_3 & = & 6.
\end{array}$$

Finally, dividing the third equation by -1 and using it to eliminate x_3 from the other equations, we arrive at

$$\begin{array}{rrrrr}
x_1 & & & = & -3 \\
& x_2 & & = & 2 \\
& & x_3 & = & 1 \\
& & 0 & = & 13.
\end{array}$$

36

The last equation is obviously impossible, so the system has no solution. In this case the system is said to be **inconsistent**. The four equations provide too many conditions to be satisfied by the three unknowns. Geometrically, there is no point which lies on all four of the planes described by the four equations. $\diamond$

Based on these examples it is tempting to say that a system has no solution if there are more equations than unknowns, and infinitely many solutions if there are fewer equations than unknowns. However, this is *not* always the case.

Example 5.4. Consider the system

$$\begin{array}{rrrrrrr} x_1 & + & x_2 & - & x_3 & = & 3 \\ 2x_1 & + & 3x_2 & + & 2x_3 & = & 4 \\ 2x_1 & + & x_2 & - & 6x_3 & = & 8 \\ 3x_1 & + & 5x_2 & + & 5x_3 & = & 5. \end{array}$$

Using the first equation to eliminate x_1 results in

$$\begin{array}{rrrrrrr} x_1 & + & x_2 & - & x_3 & = & 3 \\ & & x_2 & + & 4x_3 & = & -2 \\ & & -x_2 & - & 4x_3 & = & 2 \\ & & 2x_2 & + & 8x_3 & = & -4. \end{array}$$

Notice that the last three equations are multiples of one another. So when the second equation is used to eliminate x_2 we get

$$\begin{array}{rrrrr} x_1 & & - & 5x_3 & = & 5 \\ & x_2 & + & 4x_3 & = & -2 \\ & & & 0 & = & 0 \\ & & & 0 & = & 0. \end{array}$$

The final two equations are trivial, and solving for x_1 and x_2 in terms of x_3, we find

$$\begin{bmatrix} x_1 \\ x_2 \\ x_3 \end{bmatrix} = \begin{bmatrix} 5 + 5x_3 \\ -2 - 4x_3 \\ x_3 \end{bmatrix} = \begin{bmatrix} 5 \\ -2 \\ 0 \end{bmatrix} + x_3 \begin{bmatrix} 5 \\ -4 \\ 1 \end{bmatrix}.$$

Geometrically, the set of solutions is the line of intersection of the four planes described by the four equations.

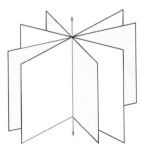

Thus the set of solution is infinite, even though there are more equations than unknowns. $\diamond$

Example 5.5. The system

$$
\begin{aligned}
x_1 + 3x_2 - 4x_3 &= 3 \\
2x_1 + 6x_2 - 8x_3 &= 2
\end{aligned}
$$

reduces to

$$
\begin{aligned}
x_1 + 3x_2 - 4x_3 &= 3 \\
0 &= -4
\end{aligned}
$$

so there are no solutions even though there are fewer equations than unknowns. Geometrically, the two planes described by these equations are parallel, and therefore do not intersect.

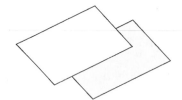

$\diamond$

Exercises

In Exercises 1 through 8 find all solutions to each system. Describe the intersection of the planes geometrically.

5.1.
$$
\begin{aligned}
3x_1 + 2x_2 - 2x_3 &= -8 \\
-3x_1 \quad\quad\ + 2x_3 &= 10 \\
3x_1 + 4x_2 - 4x_3 &= 10.
\end{aligned}
$$

5.2.
$$
\begin{aligned}
7x_1 + x_2 - 5x_3 &= 10 \\
4x_1 + x_2 - 2x_3 &= 7 \\
6x_1 + x_2 - 4x_3 &= 9.
\end{aligned}
$$

5.3.
$$
\begin{aligned}
2x_1 + 3x_2 + 9x_3 &= 3 \\
-2x_1 + x_2 + 3x_3 &= -5 \\
3x_1 + 2x_2 + 6x_3 &= 6.
\end{aligned}
$$

5.4.
$$
\begin{aligned}
-3x_1 + x_2 + 4x_3 &= 2 \\
6x_1 - 2x_2 - 8x_3 &= 1.
\end{aligned}
$$

5.5.
$$
\begin{aligned}
2x_1 + 3x_2 - x_3 &= 0 \\
-x_1 + 4x_2 + 6x_3 &= 2.
\end{aligned}
$$

5.6.
$$
\begin{aligned}
-4x_1 + 4x_2 + x_3 &= -1 \\
2x_1 - 2x_2 - x_3 &= 3 \\
8x_1 + 6x_2 + 2x_3 &= -5.
\end{aligned}
$$

5.7. $\quad \begin{aligned} 2x_1 &- x_2 + 3x_3 = -15 \\ -3x_1 &- 4x_2 + 2x_3 = -7 \\ x_1 &+ 4x_2 + 2x_3 = -3 \\ -2x_1 &+ x_2 + x_3 = 4. \end{aligned}$

5.8. $\quad \begin{aligned} x_1 & + x_3 = -1 \\ x_1 &+ x_2 + 3x_3 = -1 \\ & x_2 + 2x_3 = 0 \\ -2x_1 &+ x_2 = 4. \end{aligned}$

In Exercises 9 through 16 write the vector $\mathbf{v}$ as a linear combination of vectors in the given set, or show that $\mathbf{v}$ is not in the span of the set.

5.9. $\left\{ \begin{bmatrix} 2 \\ 1 \end{bmatrix}, \begin{bmatrix} 3 \\ 2 \end{bmatrix} \right\} \quad \mathbf{v} = \begin{bmatrix} 1 \\ 2 \end{bmatrix}$

5.10. $\left\{ \begin{bmatrix} 2 \\ 1 \\ 1 \end{bmatrix}, \begin{bmatrix} 5 \\ 8 \\ 0 \end{bmatrix}, \begin{bmatrix} 1 \\ 6 \\ -2 \end{bmatrix} \right\} \quad \mathbf{v} = \begin{bmatrix} 8 \\ -5 \\ -11 \end{bmatrix}$

5.11. $\left\{ \begin{bmatrix} 2 \\ 1 \\ 1 \end{bmatrix}, \begin{bmatrix} 5 \\ 8 \\ 0 \end{bmatrix}, \begin{bmatrix} 11 \\ 2 \\ -12 \end{bmatrix} \right\} \quad \mathbf{v} = \begin{bmatrix} 8 \\ -5 \\ -11 \end{bmatrix}$

5.12. $\left\{ \begin{bmatrix} 1 \\ 1 \\ 1 \\ 1 \end{bmatrix}, \begin{bmatrix} 1 \\ 2 \\ 3 \\ 4 \end{bmatrix}, \begin{bmatrix} 4 \\ 3 \\ 2 \\ 1 \end{bmatrix} \right\} \quad \mathbf{v} = \begin{bmatrix} 1 \\ 5 \\ 9 \\ 13 \end{bmatrix}$

5.13. $\left\{ \begin{bmatrix} 2 \\ -1 \\ 1 \end{bmatrix}, \begin{bmatrix} 5 \\ 3 \\ -1 \end{bmatrix}, \begin{bmatrix} 1 \\ 5 \\ 3 \end{bmatrix} \right\} \quad \mathbf{v} = \begin{bmatrix} -2 \\ 1 \\ -3 \end{bmatrix}$

5.14. $\left\{ \begin{bmatrix} 1 \\ 2 \\ 3 \\ 4 \end{bmatrix}, \begin{bmatrix} -2 \\ 1 \\ -1 \\ 3 \end{bmatrix} \right\} \quad \mathbf{v} = \begin{bmatrix} -1 \\ 8 \\ 7 \\ 16 \end{bmatrix}$

5.15. $\left\{ \begin{bmatrix} 1 \\ 1 \end{bmatrix}, \begin{bmatrix} 1 \\ 3 \end{bmatrix} \right\} \quad \mathbf{v} = \begin{bmatrix} 0 \\ 0 \end{bmatrix}$

5.16. $\left\{ \begin{bmatrix} 3 \\ 2 \\ -2 \end{bmatrix}, \begin{bmatrix} -1 \\ -1 \\ 2 \end{bmatrix} \right\} \quad \mathbf{v} = \begin{bmatrix} 1 \\ -1 \\ 6 \end{bmatrix}$

In Exercises 17 through 25, find all vectors which are orthogonal to all of the given vectors.

5.17. $\left\{ \begin{bmatrix} 1 \\ 1 \\ 1 \end{bmatrix}, \begin{bmatrix} 3 \\ 2 \\ 4 \end{bmatrix} \right\}$

5.18. $\left\{ \begin{bmatrix} 2 \\ -1 \\ 5 \end{bmatrix}, \begin{bmatrix} -1 \\ 0 \\ 3 \end{bmatrix} \right\}$

5.19. $\left\{ \begin{bmatrix} 1 \\ -1 \\ 1 \end{bmatrix}, \begin{bmatrix} -1 \\ 0 \\ 3 \end{bmatrix}, \begin{bmatrix} -3 \\ 2 \\ 3 \end{bmatrix} \right\}$

5.20. $\left\{ \begin{bmatrix} 1 \\ -1 \\ 1 \end{bmatrix}, \begin{bmatrix} -1 \\ 0 \\ 3 \end{bmatrix}, \begin{bmatrix} -3 \\ 2 \\ 1 \end{bmatrix} \right\}$

5.21. $\left\{ \begin{bmatrix} 2 \\ 3 \\ 5 \end{bmatrix} \right\}$

5.22. $\left\{ \begin{bmatrix} 0 \\ 0 \\ 0 \end{bmatrix} \right\}$

5.23. $\left\{ \begin{bmatrix} 1 \\ 1 \\ 1 \\ 1 \end{bmatrix}, \begin{bmatrix} 1 \\ 2 \\ 3 \\ 4 \end{bmatrix}, \begin{bmatrix} 1 \\ -1 \\ 1 \\ -1 \end{bmatrix} \right\}$

5.24. $\left\{ \begin{bmatrix} 1 \\ -1 \\ -1 \\ 1 \end{bmatrix} \right\}$

5.25. $\left\{ \begin{bmatrix} 1 \\ 0 \\ 1 \\ 0 \end{bmatrix}, \begin{bmatrix} 0 \\ 1 \\ 0 \\ 1 \end{bmatrix} \right\}$

6 Matrices

Each step in Gaussian elimination simply modifies the coefficients a_{ij} and the right hand sides b_i of the system. To save time and space it is convenient to work only with these numbers. An $m \times n$ **matrix** is a rectangular array of numbers with m rows and n columns. Given a linear system

$$
\begin{array}{ccccccccc}
a_{11}x_1 & + & a_{12}x_2 & + & \cdots & + & a_{1n}x_n & = & b_1 \\
a_{21}x_1 & + & a_{22}x_2 & + & \cdots & + & a_{2n}x_n & = & b_2 \\
\vdots & & \vdots & & & & \vdots & & \vdots \\
a_{m1}x_1 & + & a_{m2}x_2 & + & \cdots & + & a_{mn}x_n & = & b_m
\end{array}
$$

the **coefficient matrix** of the system is the $m \times n$ matrix

$$
A = \begin{bmatrix}
a_{11} & a_{12} & \cdots & a_{1n} \\
a_{21} & a_{22} & \cdots & a_{2n} \\
\vdots & \vdots & \ddots & \vdots \\
a_{m1} & a_{m2} & \cdots & a_{mn}
\end{bmatrix}
$$

whose entries are the coefficients of the system. The **augmented matrix** for the system is the $m \times (n+1)$ matrix

$$
\left[\begin{array}{cccc|c}
a_{11} & a_{12} & \cdots & a_{1n} & b_1 \\
a_{21} & a_{22} & \cdots & a_{2n} & b_2 \\
\vdots & \vdots & \ddots & \vdots & \vdots \\
a_{m1} & a_{m2} & \cdots & a_{mn} & b_m
\end{array}\right]
$$

obtained by augmenting the coefficient matrix with the entries on the right hand side of the equations. The vertical line distinguishes an augmented matrix from a coefficient matrix.

Example 6.1. The system

$$
\begin{array}{ccccccc}
3x_1 & - & 2x_2 & + & x_3 & = & 9 \\
4x_1 & & & + & 5x_3 & = & 2
\end{array}
$$

has coefficient matrix

$$
\begin{bmatrix}
3 & -2 & 1 \\
4 & 0 & 5
\end{bmatrix}
$$

and augmented matrix

$$
\left[\begin{array}{ccc|c}
3 & -2 & 1 & 9 \\
4 & 0 & 5 & 2
\end{array}\right].
$$

Notice that the absence of x_2 in the second equation simply means its coefficient is zero. $\diamond$

Each equation of a linear system corresponds to a row of its augmented matrix, and thus each step in the elimination process corresponds to a **row operation**. There are three types of row operations.

1. Divide a row by a nonzero scalar.

41

2. Subtract a scalar multiple of one row from another row.

3. Exchange two rows.

We will explain the need for the third operation momentarily. The most important fact is that *row operations do not change the set of solutions of the system.* The goal of elimination is to put the augmented matrix in **reduced row echelon form**. A matrix is in reduced row echelon form if the following conditions hold.

Properties of Reduced Row Echelon Form

1. The first nonzero entry in each row equals 1. These entries are called the **pivot** entries.

2. Each pivot is further to the right than the pivot of the row immediately above.

3. All other entries in the column of any pivot are zero.

4. If a row has all zero entries, then so does every row below.

Example 6.2. The following is a list of all possible 2×3 reduced row echelon form matrices.

$$\begin{bmatrix} 1 & 0 & * \\ 0 & 1 & * \end{bmatrix} \quad \begin{bmatrix} 1 & * & 0 \\ 0 & 0 & 1 \end{bmatrix} \quad \begin{bmatrix} 0 & 1 & 0 \\ 0 & 0 & 1 \end{bmatrix} \quad \begin{bmatrix} 1 & * & * \\ 0 & 0 & 0 \end{bmatrix} \quad \begin{bmatrix} 0 & 1 & * \\ 0 & 0 & 0 \end{bmatrix} \quad \begin{bmatrix} 0 & 0 & 1 \\ 0 & 0 & 0 \end{bmatrix} \quad \begin{bmatrix} 0 & 0 & 0 \\ 0 & 0 & 0 \end{bmatrix}$$

The asterisks can be any real numbers, including zero. $\diamond$

Two matrices are said to be **row equivalent** if one can be obtained from the other by a sequence of row operations. Since row operations are reversible, this relationship is symmetric. It turns out that every matrix is row equivalent to a unique matrix in reduced row echelon form.

Proposition 6.1. Let A be any $m \times n$ matrix. Then A is row equivalent to exactly one reduced row echelon form matrix. This matrix is called the reduced row echelon form of A and is denoted $\mathrm{rref}(A)$.

The examples in this section illustrate how one might construct an algorithm for finding the reduced row echelon form of a matrix A. The proof of its uniqueness is somewhat technical and can be found in Appendix C.

Example 6.3. Find the reduced row echelon form of

$$A = \begin{bmatrix} 1 & 1 & 0 & 0 & 2 \\ 2 & 2 & 0 & 1 & 6 \\ 0 & 1 & -1 & 1 & 3 \\ -1 & -2 & 1 & 1 & -1 \end{bmatrix}.$$

Solution. Subtracting twice the first row from the second and adding the first row to the fourth results in

$$\begin{bmatrix} 1 & 1 & 0 & 0 & 2 \\ 0 & 0 & 0 & 1 & 2 \\ 0 & 1 & -1 & 1 & 3 \\ 0 & -1 & 1 & 1 & 1 \end{bmatrix}.$$

Notice that the first nonzero entry in the third row is to the *left* of the first nonzero entry in the second row. This violates one of the conditions of echelon form. The solution is to exchange the second row with the third row.

$$\begin{bmatrix} 1 & 1 & 0 & 0 & 2 \\ 0 & 1 & -1 & 1 & 3 \\ 0 & 0 & 0 & 1 & 2 \\ 0 & -1 & 1 & 1 & 1 \end{bmatrix}$$

Now we can use the second row to eliminate the other nonzero entries in the second column. Subtracting the second row from the first and adding the second row to the fourth gives

$$\begin{bmatrix} 1 & 0 & 1 & -1 & -1 \\ 0 & 1 & -1 & 1 & 3 \\ 0 & 0 & 0 & 1 & 2 \\ 0 & 0 & 0 & 2 & 4 \end{bmatrix}.$$

The first nonzero entry in the third row is in the fourth column, so we use it to eliminate the other entries in the fourth column. Adding the third row to the first, subtracting the third row from the second, and subtracting twice the third row from the fourth we get

$$\begin{bmatrix} 1 & 0 & 1 & 0 & 1 \\ 0 & 1 & -1 & 0 & 1 \\ 0 & 0 & 0 & 1 & 2 \\ 0 & 0 & 0 & 0 & 0 \end{bmatrix},$$

which is in reduced row echelon form. $\diamond$

Given a system whose augmented matrix is in reduced row echelon form, it is very easy to determine the set of solutions. We identify the variables as one of two types. Those variables whose corresponding column contains a pivot are called **pivot variables**. The remaining variables are called **free variables**. By the properties of reduced row echelon form, every nonzero equation of the system either contains a pivot variable, or is the equation $0 = 1$. That is, any equation which actually contains a variable contains a pivot variable. Moreover, each pivot variable appears in *exactly one* equation of the system and it is *the only pivot variable* in that equation. Thus, given any choice of the free variables, we can solve for the pivot variables, and the only possible inconsistency that can arise is an equation of the form $0 = 1$.

Example 6.4. Find the set of solutions of the system

$$\begin{aligned}
x_1 + 2x_2 + x_3 + x_4 &= 7 \\
x_1 + 2x_2 + 2x_3 - x_4 &= 12 \\
2x_1 + 4x_2 + 6x_4 &= 4.
\end{aligned}$$

Solution. The augmented matrix of this system is

$$\begin{bmatrix} 1 & 2 & 1 & 1 & | & 7 \\ 1 & 2 & 2 & -1 & | & 12 \\ 2 & 4 & 0 & 6 & | & 4 \end{bmatrix}.$$

Its reduced row echelon form is

$$\begin{bmatrix} 1 & 2 & 0 & 3 & | & 2 \\ 0 & 0 & 1 & -2 & | & 5 \\ 0 & 0 & 0 & 0 & | & 0 \end{bmatrix}.$$

(Verify this!) Thus the system reduces to

$$\begin{aligned}
x_1 + 2x_2 + 3x_4 &= 2 \\
x_3 - 2x_4 &= 5 \\
0 &= 0.
\end{aligned}$$

Solving for the pivot variables x_1 and x_3 in terms of the free variables x_2 and x_4 gives $x_1 = 2 - 2x_2 - 3x_4$ and $x_3 = 5 + 2x_4$. The solution therefore takes the form

$$\begin{bmatrix} x_1 \\ x_2 \\ x_3 \\ x_4 \end{bmatrix} = \begin{bmatrix} 2 - 2x_2 - 3x_4 \\ x_2 \\ 5 + 2x_4 \\ x_4 \end{bmatrix} = \begin{bmatrix} 2 \\ 0 \\ 5 \\ 0 \end{bmatrix} + x_2 \begin{bmatrix} -2 \\ 1 \\ 0 \\ 0 \end{bmatrix} + x_4 \begin{bmatrix} -3 \\ 0 \\ 2 \\ 1 \end{bmatrix}.$$

This is a parametric representation of a plane in $\mathbf{R}^4$. $\diamond$

Example 6.5. Find the set of solutions of the system

$$\begin{aligned}
x_1 + 2x_2 + x_3 + x_4 &= 8 \\
x_1 + 2x_2 + 2x_3 - x_4 &= 12 \\
2x_1 + 4x_2 + 6x_4 &= 4.
\end{aligned}$$

Solution. Notice that this is exactly the same system as in Example 6.4 except for the right hand side of the first equation. Since

$$\text{rref} \begin{bmatrix} 1 & 2 & 1 & 1 & | & 8 \\ 1 & 2 & 2 & -1 & | & 12 \\ 2 & 4 & 0 & 6 & | & 4 \end{bmatrix} = \begin{bmatrix} 1 & 2 & 0 & 3 & | & 0 \\ 0 & 0 & 1 & -2 & | & 0 \\ 0 & 0 & 0 & 0 & | & 1 \end{bmatrix}$$

the system reduces to

$$\begin{aligned}
x_1 + 2x_2 + 3x_4 &= 0 \\
x_3 - 2x_4 &= 0 \\
0 &= 1.
\end{aligned}$$

The last equation is impossible so the system has no solutions. $\diamond$

Given a system whose augmented matrix is in reduced row echelon form, the size of the solution set is determined by two factors – the absence or presence of the equation $0 = 1$, and the number of free variables. The system has no solutions if the equation $0 = 1$ appears. If the system does not contain the equation $0 = 1$, the system is consistent. In this case, if there are no free variables, there is a unique solution. If there are free variables (one or more) then there are infinitely many solutions, one for each choice of values for the free variables. These observations are summarized in the following proposition.

Proposition 6.2. Given a system whose augmented matrix is in reduced row echelon form, there are three possibilities.

1. **No solutions.** One of the equations is $0 = 1$.

2. **Exactly one solution.** There are no free variables, and no equation $0 = 1$.

3. **Infinitely many solutions.** There is at least one free variable, and no equation $0 = 1$.

Since, by Proposition 6.1, every matrix is row equivalent to a unique reduced row echelon form matrix, and since row operations on the augmented matrix do not change the set of solutions of the system, the three possibilities listed above apply to *any* system of linear equations. So *every linear system has no solutions, exactly one solution, or infinitely many solutions.*

Exercises

In Exercises 1 through 6 express the given system as an augmented matrix, use Gaussian elimination to put the system in reduced row echelon form, and determine whether the system has no solutions, one solution (find it) or infinitely many solutions.

6.1.
$$
\begin{aligned}
w & & + & y & & & = & 5 \\
w & + 2x & + & 3y & + & 4z & = & 13 \\
w & + 2x & + & y & + & 2z & = & 5.
\end{aligned}
$$

6.2.
$$
\begin{aligned}
x & + 2y & + 3z & = 1 \\
2x & + y & - 2z & = 1.
\end{aligned}
$$

6.3.
$$
\begin{aligned}
x & - 2y & + & z & = 0 \\
2x & + 2y & - & z & = 8 \\
3x & + y & + & 2z & = 0 \\
x & - 2y & + & 3z & = -7.
\end{aligned}
$$

6.4.
$$
\begin{aligned}
w & + 2x & - & y & - & z & = 1 \\
2w & + 4x & - & 2y & + & 3z & = 3 \\
-w & + x & - & 2y & + & 4z & = 2.
\end{aligned}
$$

$$
\begin{array}{r}
6.5. \qquad u + 2v + 3w = -1 \\
u + 2v + 4w = -2 \\
-2u - 4v - 4w = 2.
\end{array}
$$

$$
\begin{array}{r}
6.6. \qquad x + y + z = 3 \\
x + 2y + 3z = 0 \\
x + 3y + 4z = -2.
\end{array}
$$

6.7. In each part, find the reduced row echelon form of the given augmented matrix. Write the system of equations corresponding to the original augmented matrix and to the reduced row echelon form matrix. Find all solutions.

(a) $\left[\begin{array}{ccc|c} 2 & 3 & 1 & 0 \\ -1 & 4 & 0 & 0 \\ -6 & 2 & -2 & 0 \end{array}\right]$

(b) $\left[\begin{array}{ccc|c} 2 & 3 & 1 & -3 \\ -1 & 4 & 0 & 1 \\ -6 & 2 & -2 & 6 \end{array}\right]$

In Exercises 8 through 12, write the vector $\mathbf{v}$ as a linear combination of the vectors in the given set, or show that $\mathbf{v}$ is not in the span.

6.8. $\left\{ \begin{bmatrix} 2 \\ 3 \\ 2 \end{bmatrix}, \begin{bmatrix} -4 \\ 1 \\ 8 \end{bmatrix}, \begin{bmatrix} 5 \\ 2 \\ 4 \end{bmatrix} \right\}$ $\qquad \mathbf{v} = \begin{bmatrix} 4 \\ 3 \\ 1 \end{bmatrix}$

6.9. $\left\{ \begin{bmatrix} 2 \\ 3 \\ 2 \end{bmatrix}, \begin{bmatrix} -4 \\ 1 \\ 8 \end{bmatrix}, \begin{bmatrix} -8 \\ 9 \\ 30 \end{bmatrix} \right\}$ $\qquad \mathbf{v} = \begin{bmatrix} 0 \\ -2 \\ 1 \end{bmatrix}$

6.10. $\left\{ \begin{bmatrix} 1 \\ -4 \\ 5 \end{bmatrix}, \begin{bmatrix} -2 \\ 3 \\ 1 \end{bmatrix}, \begin{bmatrix} 1 \\ 0 \\ 4 \end{bmatrix} \right\}$ $\qquad \mathbf{v} = \begin{bmatrix} -6 \\ 18 \\ -9 \end{bmatrix}$

6.11. $\left\{ \begin{bmatrix} 7 \\ 5 \\ -2 \end{bmatrix}, \begin{bmatrix} 2 \\ -3 \\ 1 \end{bmatrix} \right\}$ $\qquad \mathbf{v} = \begin{bmatrix} 2 \\ -2 \\ 1 \end{bmatrix}$

6.12. $\left\{ \begin{bmatrix} -4 \\ 2 \end{bmatrix}, \begin{bmatrix} 1 \\ 3 \end{bmatrix} \right\}$ $\qquad \mathbf{v} = \begin{bmatrix} -4 \\ 9 \end{bmatrix}$

6.13. Find the coefficients of the quadratic polynomial $p(x) = ax^2 + bx + c$ which passes through the points $(1, 1)$, $(2, 2)$ and $(-1, 5)$.

6.14. Is there a polynomial $p(x) = x^3 + ax^2 + bx + c$ which satisfies $p(-2) = 2$, $p(-1) = 3$, $p(1) = 0$ and $p(2) = 8$?

6.15. Suppose $f(t) = A \cos t + B \sin t + C e^t$ and $f(0) = 2$, $f'(0) = 0$, $f''(0) = 6$. Find A, B and C.

7 Matrix-Vector Products

In this section we will investigate the structure of solution sets of linear systems. First we introduce some important notation. Given a linear system

$$
\begin{array}{ccccccccc}
a_{11}x_1 & + & a_{12}x_2 & + & \cdots & + & a_{1n}x_n & = & b_1 \\
a_{21}x_1 & + & a_{22}x_2 & + & \cdots & + & a_{2n}x_n & = & b_2 \\
\vdots & & \vdots & & & & \vdots & & \vdots \\
a_{m1}x_1 & + & a_{m2}x_2 & + & \cdots & + & a_{mn}x_n & = & b_m
\end{array}
\tag{7.1}
$$

we may rewrite it as a single vector equation

$$
\begin{bmatrix}
a_{11}x_1 & + & a_{12}x_2 & + & \cdots & + & a_{1n}x_n \\
a_{21}x_1 & + & a_{22}x_2 & + & \cdots & + & a_{2n}x_n \\
\vdots & & \vdots & & & & \vdots \\
a_{m1}x_1 & + & a_{m2}x_2 & + & \cdots & + & a_{mn}x_n
\end{bmatrix}
=
\begin{bmatrix}
b_1 \\
b_2 \\
\vdots \\
b_m
\end{bmatrix}.
\tag{7.2}
$$

Let A denote the coefficient matrix of the system, and let $\mathbf{x}$ denote the vector of unknowns x_1 through x_n. Then we define the **matrix-vector product** of the matrix A with the vector $\mathbf{x}$ to be the vector

$$
A\mathbf{x} =
\begin{bmatrix}
a_{11} & a_{12} & \cdots & a_{1n} \\
a_{21} & a_{22} & \cdots & a_{2n} \\
\vdots & \vdots & & \vdots \\
a_{m1} & a_{m2} & \cdots & a_{mn}
\end{bmatrix}
\begin{bmatrix}
x_1 \\
x_2 \\
\vdots \\
x_n
\end{bmatrix}
=
\begin{bmatrix}
a_{11}x_1 & + & a_{12}x_2 & + & \cdots & + & a_{1n}x_n \\
a_{21}x_1 & + & a_{22}x_2 & + & \cdots & + & a_{2n}x_n \\
\vdots & & \vdots & & & & \vdots \\
a_{m1}x_1 & + & a_{m2}x_2 & + & \cdots & + & a_{mn}x_n
\end{bmatrix}.
$$

The system (7.1) can now be written concisely as

$$
\boxed{A\mathbf{x} = \mathbf{b}}
$$

where $\mathbf{b}$ is the vector on the right hand side of (7.2).

There are two important ways of viewing the product $A\mathbf{x}$. The first is to write $A\mathbf{x}$ as a sum of n vectors and factor out the components x_i of $\mathbf{x}$ to get

$$
A\mathbf{x} = x_1
\begin{bmatrix}
a_{11} \\
a_{21} \\
\vdots \\
a_{m1}
\end{bmatrix}
+ x_2
\begin{bmatrix}
a_{12} \\
a_{22} \\
\vdots \\
a_{m2}
\end{bmatrix}
+ \cdots + x_n
\begin{bmatrix}
a_{1n} \\
a_{2n} \\
\vdots \\
a_{mn}
\end{bmatrix},
$$

the linear combinations of the *columns* of A obtained by using the components of $\mathbf{x}$ as coefficients. So if we let $\mathbf{v}_1, \mathbf{v}_2, \ldots, \mathbf{v}_n$ denote the columns of A then

$$A\mathbf{x} = \begin{bmatrix} | & | & & | \\ \mathbf{v}_1 & \mathbf{v}_2 & \cdots & \mathbf{v}_n \\ | & | & & | \end{bmatrix} \begin{bmatrix} x_1 \\ x_2 \\ \vdots \\ x_n \end{bmatrix} = x_1\mathbf{v}_1 + x_2\mathbf{v}_2 + \cdots + x_n\mathbf{v}_n. \qquad (7.3)$$

So $A\mathbf{x}$ is the linear combination of the columns of A whose coefficients are the components of $\mathbf{x}$.

On the other hand, we can view the product in terms of the rows of A. To do this correctly we need to introduce some notation. When we speak of a vector $\mathbf{v}$ in $\mathbf{R}^n$ we always mean a **column vector**. We define the **transpose** of the column vector $\mathbf{v}$ to be the **row vector** $\mathbf{v}^T$ whose components are the same as the components of $\mathbf{v}$, but are written horizontally instead of vertically. That is,

$$\mathbf{v} = \begin{bmatrix} v_1 \\ v_2 \\ \vdots \\ v_n \end{bmatrix} \qquad \Longrightarrow \qquad \mathbf{v}^T = \begin{bmatrix} v_1 & v_2 & \cdots & v_n \end{bmatrix}.$$

Now we observe that the first component of the product $A\mathbf{x}$ is the dot product of the vector

$$\mathbf{w}_1 = \begin{bmatrix} a_{11} \\ a_{12} \\ \vdots \\ a_{1n} \end{bmatrix}$$

with $\mathbf{x}$, and that the row vector $\mathbf{w}_1^T$ is the first row of the matrix A. Likewise, component i of $A\mathbf{x}$ is the dot product of

$$\mathbf{w}_i = \begin{bmatrix} a_{i1} \\ a_{i2} \\ \vdots \\ a_{in} \end{bmatrix}$$

with $\mathbf{x}$, and $\mathbf{w}_i^T$ is row i of A. Thus

$$A\mathbf{x} = \begin{bmatrix} \text{---} & \mathbf{w}_1^T & \text{---} \\ \text{---} & \mathbf{w}_2^T & \text{---} \\ & \vdots & \\ \text{---} & \mathbf{w}_m^T & \text{---} \end{bmatrix} \begin{bmatrix} | \\ \mathbf{x} \\ | \end{bmatrix} = \begin{bmatrix} \mathbf{w}_1 \cdot \mathbf{x} \\ \mathbf{w}_2 \cdot \mathbf{x} \\ \vdots \\ \mathbf{w}_m \cdot \mathbf{x} \end{bmatrix}.$$

Example 7.1. Let

$$A = \begin{bmatrix} 1 & 2 & 3 \\ 2 & -1 & 3 \end{bmatrix} \qquad \text{and} \qquad \mathbf{x} = \begin{bmatrix} 3 \\ 2 \\ -1 \end{bmatrix}.$$

Then

$$A\mathbf{x} = 3 \begin{bmatrix} 1 \\ 2 \end{bmatrix} + 2 \begin{bmatrix} 2 \\ -1 \end{bmatrix} - 1 \begin{bmatrix} 3 \\ 3 \end{bmatrix} = \begin{bmatrix} 4 \\ 1 \end{bmatrix}$$

or, equivalently

$$A\mathbf{x} = \begin{bmatrix} 1(3) + 2(2) + 3(-1) \\ 2(3) - 1(2) + 3(-1) \end{bmatrix} = \begin{bmatrix} 4 \\ 1 \end{bmatrix}.$$

$\diamond$

Notice that the product $A\mathbf{x}$ only makes sense if the number of columns of A equals the number of components of $\mathbf{x}$, i.e. when $\mathbf{x}$ is in $\mathbf{R}^n$ and A is $m \times n$ for some m. The vector $A\mathbf{x}$ then has m components and is therefore a vector in $\mathbf{R}^m$. We can therefore think of multiplication by A as a function which takes vectors in $\mathbf{R}^n$ to vectors in $\mathbf{R}^m$.

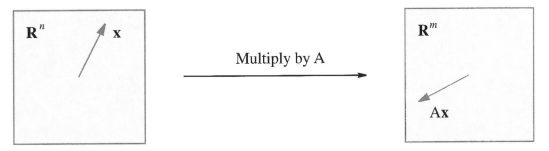

We now prove the two most important properties of matrix-vector products.

Proposition 7.1. Let A be an $m \times n$ matrix. Then

1. $A(\mathbf{x} + \mathbf{y}) = A\mathbf{x} + A\mathbf{y}$

2. $A(c\mathbf{x}) = cA\mathbf{x}$

for any vectors $\mathbf{x}$, $\mathbf{y}$ in $\mathbf{R}^n$ and any scalar $c \in \mathbf{R}$.

Proof. Let $\mathbf{v}_1$ through $\mathbf{v}_n$ denote the columns of A, let x_1 through x_n denote the components of $\mathbf{x}$ and let y_1 through y_n denote the components of $\mathbf{y}$. Then, using equation (7.3),

$$\begin{aligned} A(\mathbf{x} + \mathbf{y}) &= (x_1 + y_1)\mathbf{v}_1 + (x_2 + y_2)\mathbf{v}_2 + \cdots + (x_n + y_n)\mathbf{v}_n \\ &= (x_1\mathbf{v}_1 + x_2\mathbf{v}_2 + \cdots + x_n\mathbf{v}_n) + (y_1\mathbf{v}_1 + y_2\mathbf{v}_2 + \cdots + y_n\mathbf{v}_n) \\ &= A\mathbf{x} + A\mathbf{y} \end{aligned}$$

and

$$A(c\mathbf{x}) = (cx_1)\mathbf{v}_1 + (cx_2)\mathbf{v}_2 + \cdots + (cx_n)\mathbf{v}_n$$
$$= c(x_1\mathbf{v}_1 + x_2\mathbf{v}_2 + \cdots + x_n\mathbf{v}_n)$$
$$= cA\mathbf{x}.$$

$\square$

It follows by repeated application of these properties that

$$A(c_1\mathbf{x}_1 + c_2\mathbf{x}_2 + \cdots + c_k\mathbf{x}_k) = c_1A\mathbf{x}_1 + c_2A\mathbf{x}_2 + \cdots + c_kA\mathbf{x}_k,$$

the product of A with a linear combination of $\mathbf{x}_1$ through $\mathbf{x}_k$ equals the same linear combination of $A\mathbf{x}_1$ through $A\mathbf{x}_k$.

Exercises

In Exercises 1 through 6 calculate the matrix-vector product, or state that the product is not defined.

7.1. $A = \begin{bmatrix} 2 & 0 & 1 \\ 3 & -4 & 2 \\ 5 & 1 & -3 \end{bmatrix}$ $\qquad \mathbf{v} = \begin{bmatrix} 3 \\ -4 \\ 1 \end{bmatrix}$.

7.2. $A = \begin{bmatrix} 3 & 1 & 8 \\ -4 & 5 & 2 \end{bmatrix}$ $\qquad \mathbf{v} = \begin{bmatrix} 2 \\ -1 \\ 0 \end{bmatrix}$.

7.3. $A = \begin{bmatrix} 3 & 1 & 8 \\ -4 & 5 & 2 \end{bmatrix}$ $\qquad \mathbf{v} = \begin{bmatrix} 4 \\ -5 \end{bmatrix}$.

7.4. $A = \begin{bmatrix} 5 & -7 & 2 & 0 \\ 5 & -1 & 3 & 1 \end{bmatrix}$ $\qquad \mathbf{v} = \begin{bmatrix} -2 \\ 5 \end{bmatrix}$.

7.5. $A = \begin{bmatrix} 5 & 5 \\ -7 & -1 \\ 2 & 3 \\ 0 & 1 \end{bmatrix}$ $\qquad \mathbf{v} = \begin{bmatrix} -2 \\ 5 \end{bmatrix}$.

7.6. $A = \begin{bmatrix} 3 \\ 0 \\ -2 \\ 1 \end{bmatrix}$ $\qquad \mathbf{v} = \begin{bmatrix} 8 \\ 2 \\ 3 \\ 1 \end{bmatrix}$.

8 Null Space

Given an $m \times n$ matrix A, we say that a linear system $A\mathbf{x} = \mathbf{b}$ is **homogeneous** if $\mathbf{b} = \mathbf{0}$, and **inhomogeneous** if $\mathbf{b} \neq \mathbf{0}$. Given an $m \times n$ matrix A, the **null space** of A is

$$N(A) = \{\mathbf{x} \in \mathbf{R}^n \mid A\mathbf{x} = \mathbf{0}\},$$

the set of all solutions of the homogeneous linear system $A\mathbf{x} = \mathbf{0}$. If we think of multiplication by A as a function which sends a vector $\mathbf{x}$ in $\mathbf{R}^n$ to a vector $A\mathbf{x}$ in $\mathbf{R}^m$, the null space is then the set of vectors which are sent to the zero vector.

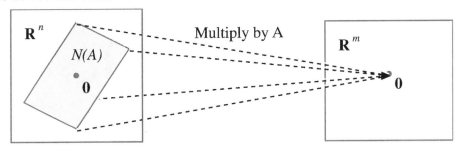

Observe that, since $A\mathbf{0} = \mathbf{0}$, the null space of A always contains at least the zero vector. So every homogeneous system is consistent. Furthermore, if $\mathbf{x}$ and $\mathbf{y}$ are solutions of $A\mathbf{x} = \mathbf{0}$, the properties of matrix-vector products imply that

$$A(\mathbf{x} + \mathbf{y}) = A\mathbf{x} + A\mathbf{y} = \mathbf{0} + \mathbf{0} = \mathbf{0}$$

and

$$A(c\mathbf{x}) = cA\mathbf{x} = c\mathbf{0} = \mathbf{0}$$

for any scalar c. Thus sums and scalar multiples of solutions of homogeneous linear systems are also solutions. These facts are summarized as follows.

Proposition 8.1. Let A be any $m \times n$ matrix. Then

1. $N(A)$ contains the zero vector.

2. If $\mathbf{x}$ and $\mathbf{y}$ are in $N(A)$, then $\mathbf{x} + \mathbf{y}$ is in $N(A)$.

3. If $\mathbf{x}$ is in $N(A)$, then $c\mathbf{x}$ is in $N(A)$ for any scalar c.

Repeated application of these properties shows that any linear combination of vectors in $N(A)$ is a vector in $N(A)$. This implies that $N(A)$ can be expressed as the span of some collection of vectors.

The null space of A is found by performing Gaussian elimination on the system $A\mathbf{x} = \mathbf{0}$. The task is simplified by observing that the final column of zeros in the augmented matrix cannot possibly become nonzero during elimination. So if we let $R = \text{rref}(A)$, the system $A\mathbf{x} = \mathbf{0}$ reduces to $R\mathbf{x} = \mathbf{0}$. So

$$N(A) = N(\text{rref}(A)),$$

the null space of A equals the null space of $\text{rref}(A)$.

Example 8.1. Let

$$A = \begin{bmatrix} 1 & 1 & 1 & 1 \\ 1 & 2 & 3 & 4 \\ 4 & 3 & 2 & 1 \end{bmatrix}.$$

Since

$$\text{rref}(A) = \begin{bmatrix} 1 & 0 & -1 & -2 \\ 0 & 1 & 2 & 3 \\ 0 & 0 & 0 & 0 \end{bmatrix}$$

the system $A\mathbf{x} = \mathbf{0}$ reduces to

$$\begin{array}{rrrrrl} x_1 & & - & x_3 & - & 2x_4 & = & 0 \\ & x_2 & + & 2x_3 & + & 3x_4 & = & 0. \end{array}$$

Solving for the pivot variables in terms of the free variables gives $x_1 = x_3 + 2x_4$ and $x_2 = -2x_3 - 3x_4$, so

$$\begin{bmatrix} x_1 \\ x_2 \\ x_3 \\ x_4 \end{bmatrix} = \begin{bmatrix} x_3 + 2x_4 \\ -2x_3 - 3x_4 \\ x_3 \\ x_4 \end{bmatrix} = x_3 \begin{bmatrix} 1 \\ -2 \\ 1 \\ 0 \end{bmatrix} + x_4 \begin{bmatrix} 2 \\ -3 \\ 0 \\ 1 \end{bmatrix}.$$

Thus

$$N(A) = \text{span}\left(\begin{bmatrix} 1 \\ -2 \\ 1 \\ 0 \end{bmatrix}, \begin{bmatrix} 2 \\ -3 \\ 0 \\ 1 \end{bmatrix} \right).$$

$\diamond$

Next let us consider an inhomogeneous system

$$A\mathbf{x} = \mathbf{b} \qquad \mathbf{b} \neq \mathbf{0}.$$

Suppose $\mathbf{x}$ and $\mathbf{y}$ are solutions of this system. Then

$$A(\mathbf{x} + \mathbf{y}) = A\mathbf{x} + A\mathbf{y} = \mathbf{b} + \mathbf{b} = 2\mathbf{b} \neq \mathbf{b}$$

so $\mathbf{x} + \mathbf{y}$ is not a solution. Likewise, for any scalar $c \neq 1$,

$$A(c\mathbf{x}) = cA\mathbf{x} = c\mathbf{b} \neq \mathbf{b}$$

so $c\mathbf{x}$ is not a solution. Thus the set of solutions of any inhomogeneous system is *not* closed under addition or scalar multiplication, and cannot be expressed as a span. However, there is a very nice relationship between the set of solutions of an inhomogeneous system $A\mathbf{x} = \mathbf{b}$ and the set of solutions of the corresponding homogeneous system $A\mathbf{x} = \mathbf{0}$.

Proposition 8.2. Suppose $\mathbf{x}_p$ is any particular solution of the inhomogeneous system $A\mathbf{x} = \mathbf{b}$. Then the set of solutions of the system $A\mathbf{x} = \mathbf{b}$ consists of all vectors of the form

$$\mathbf{x}_p + \mathbf{x}_h$$

where $\mathbf{x}_h$ is a solution of $A\mathbf{x} = \mathbf{0}$.

Proof. First we show that every vector of the form $\mathbf{x}_p + \mathbf{x}_h$ is a solution of $A\mathbf{x} = \mathbf{b}$. By the first property of matrix-vector products,

$$A(\mathbf{x}_p + \mathbf{x}_h) = A\mathbf{x}_p + A\mathbf{x}_h = \mathbf{b} + \mathbf{0} = \mathbf{b}$$

so $\mathbf{x}_p + \mathbf{x}_h$ is a solution of $A\mathbf{x} = \mathbf{b}$. Next we show that every solution of $A\mathbf{x} = \mathbf{b}$ equals $\mathbf{x}_p + \mathbf{x}_h$ for some solution $\mathbf{x}_h$ of $A\mathbf{x} = \mathbf{0}$. Suppose $\mathbf{x}$ is any solution of $A\mathbf{x} = \mathbf{b}$. Then

$$A(\mathbf{x} - \mathbf{x}_p) = A\mathbf{x} - A\mathbf{x}_p = \mathbf{b} - \mathbf{b} = \mathbf{0}$$

so $\mathbf{x} - \mathbf{x}_p$ is a solution of $A\mathbf{x} = \mathbf{0}$. Call this solution $\mathbf{x}_h$. Then $\mathbf{x} - \mathbf{x}_p = \mathbf{x}_h$, so $\mathbf{x} = \mathbf{x}_p + \mathbf{x}_h$. $\square$

Geometrically this means that the entire set of solutions of $A\mathbf{x} = \mathbf{b}$ is obtained by translating the set of solutions of $A\mathbf{x} = \mathbf{0}$ (the null space of A) by the fixed vector $\mathbf{x}_p$.

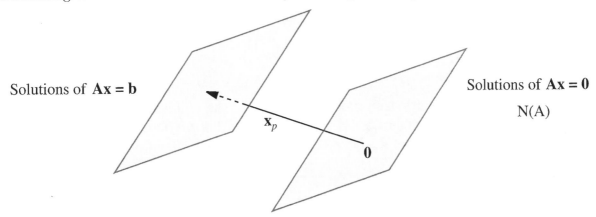

Example 8.2. Let

$$A = \begin{bmatrix} 1 & -3 \\ -1 & 3 \end{bmatrix} \quad \text{and} \quad \mathbf{b} = \begin{bmatrix} -4 \\ 4 \end{bmatrix}.$$

The system $A\mathbf{x} = \mathbf{b}$ has solution

$$\begin{bmatrix} x_1 \\ x_2 \end{bmatrix} = \begin{bmatrix} -4 \\ 0 \end{bmatrix} + x_2 \begin{bmatrix} 3 \\ 1 \end{bmatrix}$$

while the system $A\mathbf{x} = \mathbf{0}$ has solution

$$\begin{bmatrix} x_1 \\ x_2 \end{bmatrix} = x_2 \begin{bmatrix} 3 \\ 1 \end{bmatrix}.$$

Thus in this case the solution sets are parallel lines.

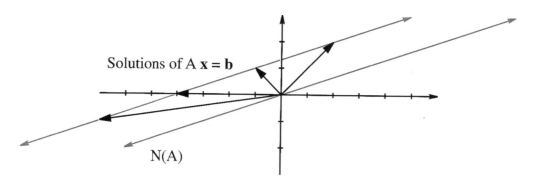

Solutions of A **x** = **b**

N(A)

The set of solutions of $A\mathbf{x} = \mathbf{b}$ is obtained by translating the null space of A by *any* solution of $A\mathbf{x} = \mathbf{b}$. ◇

This relationship between solutions of homogeneous and inhomogeneous equations is dependent upon the existence of at least one solution of the inhomogeneous equation. While a homogeneous equation $A\mathbf{x} = \mathbf{0}$ always has at least one solution, namely $\mathbf{x} = \mathbf{0}$, the inhomogeneous equation $A\mathbf{x} = \mathbf{b}$ may or may not have solutions.

Example 8.3. Let

$$A = \begin{bmatrix} 1 & -3 \\ -1 & 3 \end{bmatrix} \quad \text{and} \quad \mathbf{b} = \begin{bmatrix} -3 \\ 4 \end{bmatrix}.$$

The system $A\mathbf{x} = \mathbf{b}$ reduces to

$$\begin{aligned} x_1 \quad - \quad 3x_2 &= 0 \\ 0 &= 1 \end{aligned}$$

and therefore has no solutions. ◇

There is an important relationship between vectors in the null space of a matrix A and the columns of A. Recall that $A\mathbf{x}$ is the linear combination of the columns of A whose coefficients are the components of $\mathbf{x}$. Thus, if $A\mathbf{x} = \mathbf{0}$ we have

$$x_1\mathbf{v}_1 + x_2\mathbf{v}_2 + \cdots + x_n\mathbf{v}_n = \mathbf{0}$$

where $\mathbf{v}_1$ through $\mathbf{v}_n$ are the columns of A. Thus each element of $N(A)$ gives rise to a linear combination of the columns of A which equals the zero vector. The null space always contains the zero vector, which corresponds to the trivial combination. If the null space of A contains any nonzero vectors then the columns of A are linearly dependent.

Example 8.4. Let

$$\mathbf{v}_1 = \begin{bmatrix} 1 \\ 1 \\ 4 \end{bmatrix}, \quad \mathbf{v}_2 = \begin{bmatrix} 1 \\ 2 \\ 3 \end{bmatrix}, \quad \mathbf{v}_3 = \begin{bmatrix} 1 \\ 3 \\ 2 \end{bmatrix}, \quad \mathbf{v}_4 = \begin{bmatrix} 1 \\ 4 \\ 1 \end{bmatrix}.$$

These are the columns of the matrix A in Example 8.1. Since $N(A)$ contains nonzero vectors, the set $\{\mathbf{v}_1, \mathbf{v}_2, \mathbf{v}_3, \mathbf{v}_4\}$ is linearly dependent. In fact, from the vectors which span $N(A)$ we see that

$$\mathbf{v}_1 - 2\mathbf{v}_2 + \mathbf{v}_3 = \mathbf{0}$$

and

$$2\mathbf{v}_1 - 3\mathbf{v}_2 + \mathbf{v}_4 = \mathbf{0}.$$

$\diamondsuit$

The columns of A are linearly independent if and only if $N(A)$ contains *only* the zero vector. That is, the zero vector is the *only* solution of $A\mathbf{x} = \mathbf{0}$. By Proposition 6.2, this is the case only when the reduced system $R\mathbf{x} = \mathbf{0}$ has no free variables ($R = \mathrm{rref}(A)$). So every variable must be a pivot variable, and since each variable corresponds to a column of the coefficient matrix, this is the case if and only if each column of $\mathrm{rref}(A)$ contains a pivot. We summarize these conclusions as follows.

Proposition 8.3. Let A be an $m \times n$ matrix. Then the following statements are equivalent. (This means that if one statement holds then so do the others, and if one statement fails then so do the others.)

1. The columns of A are linearly independent.

2. $N(A) = \{\mathbf{0}\}$.

3. $\mathrm{rref}(A)$ has a pivot in each column.

One case in which the third statement necessarily fails is when A has more columns than rows. By definition there cannot be more than one pivot in any one row of $\mathrm{rref}(A)$, so if there are fewer rows than columns, then there cannot be a pivot in every column of $\mathrm{rref}(A)$. So by Proposition 8.3, the columns of A are linearly dependent. Since the columns of an $m \times n$ matrix form a collection of n vectors in $\mathbf{R}^m$, we can rephrase this result as follows.

Proposition 8.4. Any set of n vectors in $\mathbf{R}^m$ with $n > m$ is linearly dependent.

Exercises

For Exercises 1 through 9, express the null space of the given matrix A as a span of linearly independent vectors or show that $N(A) = \{\mathbf{0}\}$.

8.1. $A = \begin{bmatrix} 1 & 8 & 3 \\ -1 & -6 & -7 \\ 1 & 2 & 15 \\ -1 & -4 & -11 \end{bmatrix}$

8.2. $A = \begin{bmatrix} 0 & 1 & 0 \\ 0 & 0 & 0 \\ 0 & 0 & 0 \end{bmatrix}$

8.3. $A = \begin{bmatrix} 1 & 3 & -1 & 9 \\ 1 & 1 & 3 & 1 \\ 2 & 7 & -4 & 22 \end{bmatrix}$

8.4. $A = \begin{bmatrix} 0 & 1 & 1 \\ 0 & 0 & 1 \\ 0 & 0 & 0 \end{bmatrix}$

8.5. $A = \begin{bmatrix} -2 & 4 & -6 & 14 \\ 4 & 2 & 1 & 13 \\ 1 & -1 & -4 & 3 \\ 3 & -2 & 4 & -4 \end{bmatrix}$

8.6. $A = \begin{bmatrix} 2 & 3 & -1 \\ -4 & -6 & 2 \\ 8 & 12 & -4 \end{bmatrix}$

8.7. $A = \begin{bmatrix} 3 & 2 \\ 1 & -1 \\ 4 & 0 \\ 2 & 1 \end{bmatrix}$

8.8. $A = \begin{bmatrix} 3 & 2 & 1 \\ 0 & 2 & 1 \\ 0 & 0 & 1 \end{bmatrix}$

8.9. $A = \begin{bmatrix} 1 & -2 & 0 & 2 \\ 1 & 1 & 3 & -1 \end{bmatrix}$

8.10. In each part, express in parametric form the set of *all* solutions $\mathbf{x}$ of $A\mathbf{x} = A\mathbf{c}$ where A is the matrix from Exercise 9. Hint: $A\mathbf{x} = A\mathbf{c}$ is the same as $A(\mathbf{x} - \mathbf{c}) = \mathbf{0}$.

(a) $\mathbf{c} = \begin{bmatrix} 1 \\ 2 \\ 3 \\ 7 \end{bmatrix}$ (b) $\mathbf{c} = \begin{bmatrix} -3 \\ 2 \\ 4 \\ 2 \end{bmatrix}$ (c) $\mathbf{c} = \begin{bmatrix} \pi/6 \\ \sqrt{e} \\ \sqrt[3]{2} \\ 195 \end{bmatrix}$

8.11. Express in parametric form the set of solutions to the equation

$$\begin{bmatrix} 1 & 8 & -3 \\ -1 & -6 & -7 \\ 1 & 2 & 15 \\ -1 & -4 & -11 \end{bmatrix} \begin{bmatrix} x_1 \\ x_2 \\ x_3 \end{bmatrix} = \begin{bmatrix} 3 \\ 2 \\ 0 \\ 1 \end{bmatrix}.$$

8.12. Express in parametric form the set of solutions of the equation

$$\begin{bmatrix} 1 & -2 & 0 & 2 \\ 1 & 1 & 3 & -1 \end{bmatrix} \begin{bmatrix} x_1 \\ x_2 \\ x_3 \\ x_4 \end{bmatrix} = \begin{bmatrix} 5 \\ 11 \end{bmatrix}.$$

8.13. Express in parametric form the set of solutions to the equation

$$\begin{bmatrix} 1 & 3 & -1 & 9 \\ 1 & 1 & 3 & 1 \\ 2 & 7 & -4 & 22 \end{bmatrix} \begin{bmatrix} x_1 \\ x_2 \\ x_3 \\ x_4 \end{bmatrix} = \begin{bmatrix} 7 \\ 9 \\ 13 \end{bmatrix}.$$

In Exercises 14 through 17, (a) express $N(A)$ as a span of linearly independent vectors or show that $N(A) = \{\mathbf{0}\}$, (b) graph $N(A)$, (c) find *all* solutions of $A\mathbf{x} = \mathbf{b}$ and (d) graph the set of solutions of $A\mathbf{x} = \mathbf{b}$.

8.14. $A = \begin{bmatrix} 2 & 4 \\ -1 & -2 \end{bmatrix} \quad \mathbf{b} = \begin{bmatrix} 6 \\ -3 \end{bmatrix}.$

8.15. $A = \begin{bmatrix} 3 & 6 \\ 1 & 4 \end{bmatrix} \quad \mathbf{b} = \begin{bmatrix} 2 \\ 5 \end{bmatrix}.$

8.16. $A = \begin{bmatrix} 3 & -3 & -6 \\ 2 & -1 & -2 \\ 3 & 2 & 4 \end{bmatrix} \quad \mathbf{b} = \begin{bmatrix} 3 \\ 4 \\ 13 \end{bmatrix}.$

8.17. $A = \begin{bmatrix} 1 & 3 & -4 \\ -3 & -9 & 6 \\ 2 & 6 & -3 \end{bmatrix} \quad \mathbf{b} = \begin{bmatrix} -6 \\ 12 \\ -7 \end{bmatrix}.$

In Exercises 18 through 23 find a matrix A such that $N(A)$ is the given set S, or explain why no such matrix can exist.

8.18. $S = \text{span}\left(\begin{bmatrix} 1 \\ 0 \end{bmatrix}\right)$

8.19. $S = \left\{\begin{bmatrix} 0 \\ 0 \end{bmatrix}\right\}$

8.20. $S = \left\{\begin{bmatrix} 1 \\ 0 \end{bmatrix}\right\}$

8.21. $S = \mathbf{R}^2$

8.22. $S = \text{span}\left(\begin{bmatrix} 2 \\ 3 \end{bmatrix}\right)$

8.23. $S = \text{span}\left(\begin{bmatrix} 1 \\ 0 \end{bmatrix}\right) \cup \text{span}\left(\begin{bmatrix} 2 \\ 3 \end{bmatrix}\right)$

In Exercises 24 through 29, determine whether or not the given set of vectors is linearly independent. If linearly dependent, express one of the vectors as a linear combination of the others.

8.24. $\left\{ \begin{bmatrix} 1 \\ 4 \\ 2 \end{bmatrix}, \begin{bmatrix} -2 \\ 4 \\ 2 \end{bmatrix}, \begin{bmatrix} 3 \\ 1 \\ -1 \end{bmatrix}, \begin{bmatrix} 2 \\ 2 \\ 3 \end{bmatrix} \right\}$

8.25. $\left\{ \begin{bmatrix} 4 \\ 1 \\ 3 \end{bmatrix}, \begin{bmatrix} 1 \\ 1 \\ 1 \end{bmatrix}, \begin{bmatrix} 1 \\ -1 \\ 1 \end{bmatrix} \right\}$

8.26. $\left\{ \begin{bmatrix} 2 \\ 1 \\ 2 \end{bmatrix}, \begin{bmatrix} 4 \\ 2 \\ 9 \end{bmatrix}, \begin{bmatrix} -2 \\ -1 \\ 3 \end{bmatrix} \right\}$

8.27. $\left\{ \begin{bmatrix} 1 \\ 4 \\ 2 \\ 1 \end{bmatrix}, \begin{bmatrix} 0 \\ 1 \\ -2 \\ 3 \end{bmatrix}, \begin{bmatrix} 2 \\ 5 \\ 10 \\ -7 \end{bmatrix} \right\}$

8.28. $\left\{ \begin{bmatrix} -1 \\ 4 \\ 2 \\ 1 \end{bmatrix}, \begin{bmatrix} -2 \\ 1 \\ -2 \\ 3 \end{bmatrix}, \begin{bmatrix} 5 \\ 5 \\ 10 \\ -7 \end{bmatrix} \right\}$

8.29. $\left\{ \begin{bmatrix} 1 \\ 2 \\ 3 \\ 4 \end{bmatrix}, \begin{bmatrix} 5 \\ 6 \\ 7 \\ 8 \end{bmatrix}, \begin{bmatrix} 9 \\ 10 \\ 11 \\ 12 \end{bmatrix}, \begin{bmatrix} 13 \\ 14 \\ 15 \\ 16 \end{bmatrix} \right\}$

9 Column Space

Given an $m \times n$ matrix A, we would like to know for which vectors $\mathbf{b}$ in $\mathbf{R}^m$ the system $A\mathbf{x} = \mathbf{b}$ has a solution. First write

$$A = \begin{bmatrix} | & | & & | \\ \mathbf{v}_1 & \mathbf{v}_2 & \cdots & \mathbf{v}_n \\ | & | & & | \end{bmatrix}$$

and recall that

$$A\mathbf{x} = x_1\mathbf{v}_1 + x_2\mathbf{v}_2 + \cdots + x_n\mathbf{v}_n$$

is the linear combination of the columns of A whose coefficients are the components of $\mathbf{x}$. Thus the system $A\mathbf{x} = \mathbf{b}$ has a solution if and only if $\mathbf{b}$ can be written as a linear combination of the columns of A. Motivated by this fact, we define the **column space** of A to be

$$C(A) = \text{span}(\mathbf{v}_1, \mathbf{v}_2, \ldots, \mathbf{v}_n),$$

the span of the columns of A. Since every such linear combination takes the form $A\mathbf{x}$ for some $\mathbf{x}$ in $\mathbf{R}^n$ and since conversely every vector of the form $A\mathbf{x}$ is such a linear combination we can also express the column space as

$$C(A) = \{A\mathbf{x} \mid \mathbf{x} \in \mathbf{R}^n\}.$$

Since the columns of A are vectors in $\mathbf{R}^m$, or equivalently, since $A\mathbf{x}$ is in $\mathbf{R}^m$ for every $\mathbf{x}$ in $\mathbf{R}^n$, the column space of A is a subset of $\mathbf{R}^m$. The following fact is an immediate consequence of the definition.

Proposition 9.1. The system $A\mathbf{x} = \mathbf{b}$ has a solution if and only if $\mathbf{b}$ is in $C(A)$.

Example 9.1. Let

$$A = \begin{bmatrix} 1 & 1 & 1 & 1 \\ 1 & 2 & 3 & 4 \\ 4 & 3 & 2 & 1 \end{bmatrix}.$$

Then the column space is

$$C(A) = \text{span} \left\{ \begin{bmatrix} 1 \\ 1 \\ 4 \end{bmatrix}, \begin{bmatrix} 1 \\ 2 \\ 3 \end{bmatrix}, \begin{bmatrix} 1 \\ 3 \\ 2 \end{bmatrix}, \begin{bmatrix} 1 \\ 4 \\ 1 \end{bmatrix} \right\}.$$

We know however from Example 8.4 that the third and fourth vectors are linear combinations of the first two. Since the first two are linearly independent,

$$C(A) = \text{span} \left(\begin{bmatrix} 1 \\ 1 \\ 4 \end{bmatrix}, \begin{bmatrix} 1 \\ 2 \\ 3 \end{bmatrix} \right)$$

and the column space of A is a plane passing through the origin in $\mathbf{R}^3$.

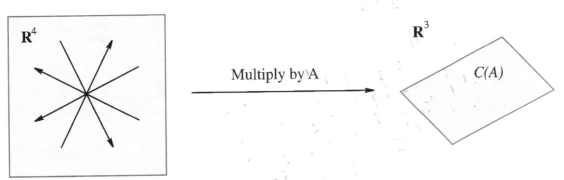

This plane is also the set of all vectors obtained by multiplying vectors in $\mathbf{R}^4$ by the matrix A. So the system $A\mathbf{x} = \mathbf{b}$ has solutions if and only if $\mathbf{b}$ is in this plane. $\diamond$

This description of the column space as a span of a set of vectors is useful in the sense that we can easily generate vectors in $C(A)$ by forming linear combinations of the spanning vectors. However, suppose we wish to know whether or not a given vector **b** is in the column space of A. We would then need to solve a system of equations. By regarding **b** as an unknown it is possible to derive conditions which determine whether or not **b** is in the column space.

Example 9.2. Let A be the matrix from Example 9.1. We will perform elimination on the system $A\mathbf{x} = \mathbf{b}$ with **b** left as an unknown. The augmented matrix for this system is

$$\left[\begin{array}{cccc|c} 1 & 1 & 1 & 1 & b_1 \\ 1 & 2 & 3 & 4 & b_2 \\ 4 & 3 & 2 & 1 & b_3 \end{array}\right].$$

Performing the row operations which place the coefficient matrix A in reduced row echelon form gives

$$\left[\begin{array}{cccc|c} 1 & 0 & -1 & -2 & 2b_1 - b_2 \\ 0 & 1 & 2 & 3 & b_2 - b_1 \\ 0 & 0 & 0 & 0 & -5b_1 + b_2 + b_3 \end{array}\right].$$

The only way an inconsistency can arise in this system is if $-5b_1 + b_2 + b_3 \neq 0$. Otherwise, the system has at least one solution, in which case **b** is in the column space of A. So **b** is in the column space of A if and only if

$$-5b_1 + b_2 + b_3 = 0.$$

This is the equation of the plane in $\mathbf{R}^3$ spanned by the columns of A. If

$$\mathbf{b}_1 = \begin{bmatrix} 2 \\ 7 \\ 3 \end{bmatrix} \qquad \text{and} \qquad \mathbf{b}_2 = \begin{bmatrix} 2 \\ 6 \\ 3 \end{bmatrix}.$$

we see that $\mathbf{b}_1$ is in $C(A)$ since $-5(2)+7+3 = 0$, but $\mathbf{b}_2$ is not in $C(A)$ since $-5(2)+6+3 \neq 0$. Thus $A\mathbf{x} = \mathbf{b}_1$ has solutions, while $A\mathbf{x} = \mathbf{b}_2$ does not. $\diamond$

In general, each row of zeros in $\text{rref}(A)$ leads to an equation that **b** must satisfy in order to be in $C(A)$.

Example 9.3. Let

$$A = \begin{bmatrix} 1 & 3 & 4 \\ 2 & 7 & 9 \\ -1 & 1 & 0 \\ 3 & 3 & 6 \end{bmatrix}.$$

Performing elimination on the augmented matrix for $A\mathbf{x} = \mathbf{b}$, we get

$$\left[\begin{array}{ccc|c} 1 & 0 & 1 & 7b_1 - 3b_2 \\ 0 & 1 & 1 & b_2 - 2b_1 \\ 0 & 0 & 0 & b_3 - 4b_2 + 9b_1 \\ 0 & 0 & 0 & b_4 + 6b_2 - 15b_1 \end{array}\right].$$

The system is consistent if and only if

$$\begin{array}{rcrcrcrcl} 9b_1 & - & 4b_2 & + & b_3 & & & = & 0 \\ -15b_1 & + & 6b_2 & & & + & b_4 & = & 0 \end{array}$$

so $\mathbf{b} \in C(A)$ if and only if $\mathbf{b}$ satisfies both equations. If we let

$$B = \begin{bmatrix} 9 & -4 & 1 & 0 \\ -15 & 6 & 0 & 1 \end{bmatrix}$$

this means that $\mathbf{b}$ satisfies $B\mathbf{b} = \mathbf{0}$, so $\mathbf{b}$ is in the null space of B. In other words $C(A) = N(B)$.

$\diamond$

If $\operatorname{rref}(A)$ does not contain a row of zeros then the system $A\mathbf{b} = \mathbf{b}$ is consistent for all $\mathbf{b}$ in $\mathbf{R}^m$.

Example 9.4. Let

$$A = \begin{bmatrix} 1 & 2 & 3 \\ 2 & 3 & 1 \end{bmatrix}.$$

Row reducing $A\mathbf{x} = \mathbf{b}$ gives

$$\left[\begin{array}{ccc|c} 1 & 0 & -7 & -3b_1 + 2b_2 \\ 0 & 1 & 5 & 2b_1 - b_2 \end{array} \right].$$

The free variable x_3 can be chosen arbitrarily and then the pivot variables are given by $x_1 = 7x_3 - 3b_1 + 2b_2$ and $x_2 = -5x_3 + 2b_1 - b_2$. There is no inconsistency, so every vector $\mathbf{b}$ in $\mathbf{R}^2$ is in the column space of A.

$\diamond$

Proposition 9.2. Let A be an $m \times n$ matrix. Then the following statements are equivalent.

1. The columns of A span $\mathbf{R}^m$.

2. $C(A) = \mathbf{R}^m$.

3. $\operatorname{rref}(A)$ has a pivot in each row.

Proof. The equivalence of 1 and 2 follows from the definition of $C(A)$, so we just need to show that 2 and 3 are equivalent. Suppose 3 holds, and let $\mathbf{b}$ be any vector in $\mathbf{R}^m$. The augmented matrix for the system $A\mathbf{x} = \mathbf{b}$ is $[A \mid \mathbf{b}]$, and the reduced row echelon form of this matrix is $[R \mid \mathbf{c}]$ for some vector $\mathbf{c}$, where $R = \operatorname{rref}(A)$. By Proposition 6.2, the reduced system $R\mathbf{x} = \mathbf{c}$ is inconsistent if and only if it contains the equation $0 = 1$. Since $\operatorname{rref}(A)$ has a pivot in each row, it does not contain a row of zeros, and the equation $0 = 1$ does not appear in the reduced system. Hence the original system $A\mathbf{x} = \mathbf{b}$ is consistent. By

Proposition 9.1 this implies $\mathbf{b}$ is in $C(A)$. Since this holds for all $\mathbf{b}$ in $\mathbf{R}^m$, 2 holds. Now suppose 3 fails, so that $\mathrm{rref}(A)$ has a row of zeros. Then, as illustrated in the examples above, this leads to an equation that must be satisfied by the components of $\mathbf{b}$ in order for $\mathbf{b}$ to be in $C(A)$. Thus not every $\mathbf{b}$ in $\mathbf{R}^m$ is in $C(A)$, so 2 fails. This proves that 2 and 3 are equivalent. $\qquad\square$

An interesting case of this proposition is the case $m = n$, where A is a **square matrix**. In this case, the presence of a pivot in each row means that there are n pivots. Since each column can contain at most one pivot, there must be a pivot in each column. The only $n \times n$ matrix in reduced row echelon form with these properties is the matrix

$$I_n = \begin{bmatrix} 1 & 0 & \cdots & 0 \\ 0 & 1 & \cdots & 0 \\ \vdots & \vdots & \ddots & \vdots \\ 0 & 0 & \cdots & 1 \end{bmatrix},$$

called the $n \times n$ **identity matrix**. Using Proposition 8.3, we reach the following conclusion.

Proposition 9.3. Let A be an $n \times n$ square matrix. Then the following statements are equivalent.

1. The columns of A span $\mathbf{R}^n$.

2. The columns of A are linearly independent.

3. $C(A) = \mathbf{R}^n$.

4. $N(A) = \{\mathbf{0}\}$.

5. $\mathrm{rref}(A) = I_n$.

So a set of n vectors in $\mathbf{R}^n$ is linearly independent if and only if it spans $\mathbf{R}^n$, and to test whether or not this is the case, we form a matrix A whose columns are these vectors and compute $\mathrm{rref}(A)$.

The next example illustrates the relationship between the column space and null space of a matrix.

Example 9.5. Let

$$A = \begin{bmatrix} 1 & -2 & 3 \\ 2 & -4 & 6 \end{bmatrix}.$$

Since the second and third columns are scalar multiples of the first column, the column space is

$$C(A) = \mathrm{span}\left(\begin{bmatrix} 1 \\ 2 \end{bmatrix} \right),$$

a line in $\mathbf{R}^2$. Since

$$\text{rref}(A) = \begin{bmatrix} 1 & -2 & 3 \\ 0 & 0 & 0 \end{bmatrix}$$

it follows that the null space is

$$N(A) = \text{span} \left(\begin{bmatrix} 2 \\ 1 \\ 0 \end{bmatrix}, \begin{bmatrix} -3 \\ 0 \\ 1 \end{bmatrix} \right),$$

a plane in $\mathbf{R}^3$. For each vector $\mathbf{b}$ in the column space of A, the system $A\mathbf{x} = \mathbf{b}$ has a solution. By Proposition 8.2, each set of solutions is a translation of the null space of A.

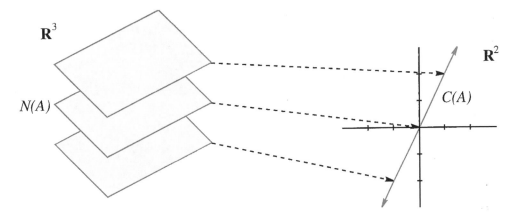

Multiplication by the matrix A sends every vector in $N(A)$ to the zero vector in $\mathbf{R}^2$. Each plane parallel to $N(A)$ is sent to a different vector in $C(A)$. ◇

Exercises

In Exercises 1 through 3 find conditions on the components of the vector $\mathbf{b}$ which are necessary and sufficient for $\mathbf{b}$ to be in the column space of the matrix A.

9.1. $\quad A = \begin{bmatrix} 1 & 8 & 3 \\ -1 & -6 & -7 \\ 1 & 2 & 15 \\ -1 & -4 & -11 \end{bmatrix} \qquad \mathbf{b} = \begin{bmatrix} b_1 \\ b_2 \\ b_3 \\ b_4 \end{bmatrix}$

9.2. $\quad A = \begin{bmatrix} 1 & 3 & -1 & 9 \\ 1 & 1 & 3 & 1 \\ 2 & 7 & -4 & 22 \end{bmatrix} \qquad \mathbf{b} = \begin{bmatrix} b_1 \\ b_2 \\ b_3 \end{bmatrix}$

9.3. $\quad A = \begin{bmatrix} 1 & 1 \\ -2 & 1 \\ 0 & 3 \\ 2 & -1 \end{bmatrix} \qquad \mathbf{b} = \begin{bmatrix} b_1 \\ b_2 \\ b_3 \\ b_4 \end{bmatrix}$

9.4. Which of the following vectors are in the column space of the matrix A from Exercise 3? Hint: Work Exercise 3 first.

(a) $\begin{bmatrix} 1 \\ 2 \\ 4 \\ 2 \end{bmatrix}$
(b) $\begin{bmatrix} 0 \\ 3 \\ 3 \\ -3 \end{bmatrix}$
(c) $\begin{bmatrix} 1 \\ 2 \\ 4 \\ -2 \end{bmatrix}$

(d) $\begin{bmatrix} -2 \\ -2 \\ -6 \\ 2 \end{bmatrix}$
(e) $\begin{bmatrix} 2 \\ 1 \\ 3 \\ -1 \end{bmatrix}$
(f) $\begin{bmatrix} 2 \\ -4 \\ 0 \\ 4 \end{bmatrix}$

In Exercises 5 through 10 find a 2×2 matrix A such that $C(A) = S$ or explain why no such matrix exists.

9.5. $S = \left\{ \begin{bmatrix} x_1 \\ x_2 \end{bmatrix} \in \mathbf{R}^2 : x_1 + x_2 = 0 \right\}$.

9.6. $S = \left\{ \begin{bmatrix} x_1 \\ x_2 \end{bmatrix} \in \mathbf{R}^2 : x_1 + x_2 = 2 \right\}$.

9.7. $S = \left\{ \begin{bmatrix} 0 \\ 0 \end{bmatrix} \right\}$.

9.8. $S = \left\{ \begin{bmatrix} 1 \\ 0 \end{bmatrix} \right\}$.

9.9. $S = \mathbf{R}^2$

9.10. $S = \text{span}\left(\begin{bmatrix} 1 \\ 0 \end{bmatrix} \right) \cup \text{span}\left(\begin{bmatrix} 0 \\ 1 \end{bmatrix} \right)$

In Exercises 11 through 19 (a) graph $N(A)$, (b) graph $C(A)$ on a separate graph, (c) find all solutions of $A\mathbf{x} = \mathbf{b}$, and (d) show that the set of solutions to $A\mathbf{x} = \mathbf{b}$ is a translate of $N(A)$.

9.11. $A = \begin{bmatrix} 1 & 3 \\ 2 & 6 \end{bmatrix}$ $\qquad \mathbf{b} = \begin{bmatrix} -2 \\ -4 \end{bmatrix}$

9.12. $A = \begin{bmatrix} 1 & 2 \\ 3 & 4 \end{bmatrix}$ $\qquad \mathbf{b} = \begin{bmatrix} 1 \\ -2 \end{bmatrix}$

9.13. $A = \begin{bmatrix} -1 & 3 & 4 \\ -2 & 6 & 8 \end{bmatrix}$ $\qquad \mathbf{b} = \begin{bmatrix} 2 \\ 4 \end{bmatrix}$

9.14. $A = \begin{bmatrix} -1 & 3 & 4 \\ -2 & 6 & 5 \end{bmatrix}$ $\qquad \mathbf{b} = \begin{bmatrix} -3 \\ 1 \end{bmatrix}$

9.15. $A = \begin{bmatrix} 1 & 2 \\ 3 & 6 \\ 2 & 4 \end{bmatrix}$ $\quad \mathbf{b} = \begin{bmatrix} 3 \\ 9 \\ 6 \end{bmatrix}$

9.16. $A = \begin{bmatrix} 1 & -2 \\ 3 & 0 \\ 2 & 4 \end{bmatrix}$ $\quad \mathbf{b} = \begin{bmatrix} 4 \\ 6 \\ 0 \end{bmatrix}$

9.17. $A = \begin{bmatrix} 1 & -2 & 3 \\ -1 & 2 & -3 \\ 2 & -4 & 6 \end{bmatrix}$ $\quad \mathbf{b} = \begin{bmatrix} -4 \\ 4 \\ -8 \end{bmatrix}$

9.18. $A = \begin{bmatrix} 2 & 1 & 3 \\ 4 & 0 & 6 \\ -2 & 1 & -3 \end{bmatrix}$ $\quad \mathbf{b} = \begin{bmatrix} 1 \\ 4 \\ -3 \end{bmatrix}$

9.19. $A = \begin{bmatrix} 2 & 3 & 5 \\ 1 & 4 & -1 \\ 0 & 2 & 5 \end{bmatrix}$ $\quad \mathbf{b} = \begin{bmatrix} -7 \\ 1 \\ -12 \end{bmatrix}$

10 Subspaces of $\mathbf{R}^n$

The null space and column space of a matrix are examples of sets called subspaces. A **linear subspace** of $\mathbf{R}^n$ is a subset V of $\mathbf{R}^n$ satisfying the following properties.

1. V contains the zero vector.

2. If $\mathbf{v}$ and $\mathbf{w}$ are in V then $\mathbf{v} + \mathbf{w}$ is in V.

3. If $\mathbf{v}$ is in V, the $c\mathbf{v}$ is in V for any scalar c.

Sets which satisfy condition 2 are said to be **closed under addition**, and sets which satisfy condition 3 are said to be **closed under scalar multiplication**. Thus subspaces are sets which contain the zero vector and are closed under both addition and scalar multiplication.

Example 10.1. The subset $\{\mathbf{0}\}$ of $\mathbf{R}^n$ consisting of just the zero vector is a subspace of $\mathbf{R}^n$, since $\mathbf{0} + \mathbf{0} = \mathbf{0}$ and $c\mathbf{0} = \mathbf{0}$ for any scalar c. This is called the **trivial** subspace. $\diamond$

Example 10.2. $\mathbf{R}^n$ is a subspace of itself. A subspace V of $\mathbf{R}^n$ is called **proper** if $V \neq \mathbf{R}^n$. $\diamond$

Example 10.3. The span of any nonempty set of vectors in $\mathbf{R}^n$ is a subspace of $\mathbf{R}^n$. To see this, suppose $V = \text{span}(\mathbf{v}_1, \mathbf{v}_2, \dots, \mathbf{v}_k)$. The trivial combination of these vectors equals the zero vector, so V contains the zero vector. If $\mathbf{v}$ and $\mathbf{w}$ are in V then

$$\mathbf{v} = a_1\mathbf{v}_1 + a_2\mathbf{v}_2 + \cdots + a_k\mathbf{v}_k$$

and

$$\mathbf{w} = b_1\mathbf{v}_1 + b_2\mathbf{v}_2 + \cdots + b_k\mathbf{v}_k.$$

So

$$\mathbf{v} + \mathbf{w} = (a_1 + b_1)\mathbf{v}_1 + (a_2 + b_2)\mathbf{v}_2 + \cdots + (a_k + b_k)\mathbf{v}_k$$

and

$$c\mathbf{v} = (ca_1)\mathbf{v}_1 + (ca_2)\mathbf{v}_2 + \cdots + (ca_k)\mathbf{v}_k$$

are both linear combinations of $\mathbf{v}_1$ through $\mathbf{v}_k$ and are therefore in V. For instance, in $\mathbf{R}^3$ any line or plane which passes through the origin is a subspace.

$\diamond$

Lines and planes which do not pass through the origin are not subspaces. The subsets in the next two examples are also not subspaces.

Example 10.4. Consider the upper half-plane

$$H = \left\{ \begin{bmatrix} x_1 \\ x_2 \end{bmatrix} \in \mathbf{R}^2 \;\middle|\; x_2 \geq 0 \right\}.$$

Suppose

$$\mathbf{v} = \begin{bmatrix} v_1 \\ v_2 \end{bmatrix} \qquad \text{and} \qquad \mathbf{w} = \begin{bmatrix} w_1 \\ w_2 \end{bmatrix}$$

are in H. Then $v_2 \geq 0$ and $w_2 \geq 0$, so

$$\mathbf{v} + \mathbf{w} = \begin{bmatrix} v_1 + w_1 \\ v_2 + w_2 \end{bmatrix}$$

is in H since $v_2 + w_2 \geq 0$. Thus H is closed under addition.

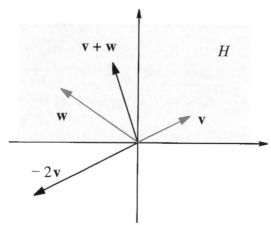

However, H is not closed under scalar multiplication. For instance, suppose $v_2 > 0$. Then

$$-2\mathbf{v} = \begin{bmatrix} -2v_1 \\ -2v_2 \end{bmatrix}$$

is not in H since $-2v_2 < 0$. So H is not a subspace. $\diamond$

Example 10.5. Let

$$\mathbf{v}_1 = \begin{bmatrix} 2 \\ 1 \end{bmatrix} \quad \text{and} \quad \mathbf{v}_2 = \begin{bmatrix} -3 \\ 1 \end{bmatrix}.$$

By Example 10.3 both $V_1 = \text{span}(\mathbf{v}_1)$ and $V_2 = \text{span}(\mathbf{v}_2)$ are subspaces. Let $V = V_1 \cup V_2$, the union of V_1 and V_2. Let $\mathbf{v}$ be in V and let c be any scalar. Then either $\mathbf{v}$ is in V_1 in which case $c\mathbf{v} \in V_1$, or $\mathbf{v}$ is in V_2 in which case $c\mathbf{v} \in V_2$. In either case $c\mathbf{v}$ is in $V_1 \cup V_2 = V$, so V is closed under scalar multiplication. However, V is not closed under addition, since for instance both $\mathbf{v}_1$ and $\mathbf{v}_2$ are in V, but

$$\mathbf{v}_1 + \mathbf{v}_2 = \begin{bmatrix} -1 \\ 2 \end{bmatrix}$$

is not in V_1 or V_2, and thus is not in V.

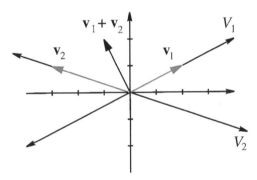

So V is not a subspace. $\diamond$

As mentioned above, the null space and column space of a matrix are subspaces.

Proposition 10.1. Let A be an m by n matrix. Then

1. The column space of A is a subspace of $\mathbf{R}^m$.

2. The null space of A is a subspace of $\mathbf{R}^n$.

Proof. The first statement follows from Example 10.3 since the column space of A is the span of the columns of A. The second statement follows immediately from Proposition 8.1. $\square$

The set of solutions of a homogeneous linear system $A\mathbf{x} = \mathbf{0}$ is a subspace, namely the null space of A. However, the set of solutions of an inhomogeneous system $A\mathbf{x} = \mathbf{b}$ (if there are solutions) is not a subspace, but rather a translation of the null space of A. We say that a subset A of $\mathbf{R}^n$ is an **affine linear subspace** if A is a translation of some subspace V of $\mathbf{R}^n$. That is, A is an affine subspace if

$$A = \{\mathbf{x}_0 + \mathbf{v} \mid \mathbf{v} \in V\}$$

for some subspace V and some fixed vector $\mathbf{x}_0$.

Example 10.6. Recall that any line in $\mathbf{R}^n$ can be described parametrically as

$$L = \{\mathbf{x}_0 + t\mathbf{v} \mid t \in \mathbf{R}\}.$$

So L is a translation of the subspace $V = \text{span}(\mathbf{v})$, and therefore an affine subspace. Likewise, any plane in $\mathbf{R}^n$ can be described as

$$P = \{\mathbf{x}_0 + s\mathbf{v}_1 + t\mathbf{v}_2 \mid t \in \mathbf{R}\}.$$

So P is a translation of the subspace $V = \text{span}(\mathbf{v}_1, \mathbf{v}_2)$, and therefore an affine subspace. In $\mathbf{R}^3$ a plane can also be described by an equation of the form

$$a_1 x_1 + a_2 x_2 + a_3 x_3 = b.$$

The set of solutions of this inhomogeneous equation are a translation of the null space of the matrix

$$A = \begin{bmatrix} a_1 & a_2 & a_3 \end{bmatrix}.$$

$\Diamond$

We emphasize that affine subspaces are *not* always subspaces. Only the ones which pass through the origin are subspaces. See the exercises.

Exercises

In Exercises 11 through 20 determine whether or not the given set is a subspace of $\mathbf{R}^n$. If it is, show that the three subspace properties are satisfied, and if it is not, show by example that one of the properties fails.

10.11. The line $3x - 2y = 2$.

10.12. The line $3x - 2y = 0$.

10.13. The set of solutions of $A\mathbf{x} = \mathbf{0}$, where A is any $m \times n$ matrix.

10.14. The set of solutions of $A\mathbf{x} = \mathbf{b}$, where A is any $m \times n$ matrix, and $\mathbf{b}$ is any vector in $\mathbf{R}^m$.

10.15. The set of solutions of $A\mathbf{x} = \mathbf{x}$, where A is any $n \times n$ matrix.

10.16. The first quadrant in $\mathbf{R}^2$. That is,

$$\left\{ \begin{bmatrix} x_1 \\ x_2 \end{bmatrix} \in \mathbf{R}^2 \;\middle|\; x_1 \geq 0 \text{ and } x_2 \geq 0 \right\}.$$

10.17. The union of the first and third quadrants in $\mathbf{R}^2$. That is,

$$\left\{ \begin{bmatrix} x_1 \\ x_2 \end{bmatrix} \in \mathbf{R}^2 \;\middle|\; (x_1 \geq 0 \text{ and } x_2 \geq 0) \text{ or } (x_1 \leq 0 \text{ and } x_2 \leq 0) \right\}.$$

10.18. The intersection of the planes $x + y - z = 3$ and $2x - y + 3z = 0$.

10.19. The intersection of the planes $x - y + 2z = 0$ and $2x + y + 4z = 0$.

10.20. The unit disk in $\mathbf{R}^2$. That is

$$\left\{ \begin{bmatrix} x_1 \\ x_2 \end{bmatrix} \in \mathbf{R} \;\middle|\; x_1^2 + x_2^2 \leq 1 \right\}.$$

10.21. Show that if V and W are subspaces of $\mathbf{R}^n$ then their intersection $V \cap W$ is also a subspace of $\mathbf{R}^n$.

10.22. Let $V + W = \{ \mathbf{v} + \mathbf{w} \mid \mathbf{v} \text{ is in } V \text{ and } \mathbf{w} \text{ is in } W \}$. Show that if V and W are subspaces then $V + W$ is a subspace.

10.23. Show that every subspace of $\mathbf{R}^2$ is one of the following: the trivial subspace $\{\mathbf{0}\}$, a line through the origin, or $\mathbf{R}^2$ itself.

10.24. Show that if A is an affine subspace of $\mathbf{R}^n$ and A contains the origin, then A is a subspace of $\mathbf{R}^n$.

11 Basis for a Subspace

A **basis** for a subspace V of $\mathbf{R}^n$ is a linearly independent set of vectors $\{\mathbf{v}_1, \mathbf{v}_2, \ldots, \mathbf{v}_k\}$ such that $V = \mathrm{span}(\mathbf{v}_1, \mathbf{v}_2 \ldots, \mathbf{v}_k)$. A basis can be thought of as an economical way of describing a subspace, in that

1. its span is the subspace – so it describes the subspace, and

2. its vectors are independent – there is no redundancy.

Given a basis for a subspace, every vector in the subspace can be described *uniquely* in terms of the basis vectors.

Proposition 11.1. Suppose $\{\mathbf{v}_1, \mathbf{v}_2, \ldots, \mathbf{v}_k\}$ is a basis for a subspace V of $\mathbf{R}^n$. Then every vector $\mathbf{v}$ in V can be expressed uniquely as a linear combination of $\mathbf{v}_1, \mathbf{v}_2, \ldots, \mathbf{v}_k$.

Proof. By definition, $\{\mathbf{v}_1, \mathbf{v}_2, \ldots, \mathbf{v}_k\}$ spans V, so every $\mathbf{v}$ in V can be written

$$\mathbf{v} = c_1\mathbf{v}_1 + c_2\mathbf{v}_2 + \cdots + c_k\mathbf{v}_k$$

for some scalars $c_1, c_2, \ldots, c_k$. To prove uniqueness, suppose in addition that

$$\mathbf{v} = d_1\mathbf{v}_1 + d_2\mathbf{v}_2 + \cdots + d_k\mathbf{v}_k.$$

Then, subtracting this from the previous equation gives

$$\mathbf{0} = (c_1 - d_1)\mathbf{v}_1 + (c_2 - d_2)\mathbf{v}_2 + \cdots + (c_k - d_k)\mathbf{v}_k.$$

Again by the definition of a basis, $\mathbf{v}_1, \mathbf{v}_2, \ldots, \mathbf{v}_k$ are linearly independent, so $c_1 - d_1 = 0$, $c_2 - d_2 = 0, \ldots, c_k - d_k = 0$. Thus $c_1 = d_1$, $c_2 = d_2$, $\ldots$, $c_k = d_k$ and the two expressions are the same. $\qquad\square$

Bases, however are not unique.

Example 11.1. The set $\{\mathbf{e}_1, \mathbf{e}_2, \ldots, \mathbf{e}_n\}$ is linearly independent and spans $\mathbf{R}^n$, and is therefore a basis for $\mathbf{R}^n$. This is called the **standard basis** for $\mathbf{R}^n$. $\qquad\diamond$

Example 11.2. Let

$$\mathbf{v}_1 = \begin{bmatrix} 1 \\ 2 \\ 4 \end{bmatrix}, \quad \mathbf{v}_2 = \begin{bmatrix} 2 \\ 1 \\ -1 \end{bmatrix}, \quad \text{and} \quad \mathbf{v}_3 = \begin{bmatrix} 3 \\ 4 \\ 5 \end{bmatrix}.$$

Let A be the matrix whose columns are $\mathbf{v}_1$, $\mathbf{v}_2$ and $\mathbf{v}_3$. Since

$$\mathrm{rref}(A) = \begin{bmatrix} 1 & 0 & 0 \\ 0 & 1 & 0 \\ 0 & 0 & 1 \end{bmatrix},$$

Proposition 9.3 implies that the columns of A are linearly independent and span $\mathbf{R}^3$. Thus $\{\mathbf{v}_1, \mathbf{v}_2, \mathbf{v}_3\}$ is a basis for $\mathbf{R}^3$. $\qquad\diamond$

By Proposition 9.3, any set of n linearly independent vectors in $\mathbf{R}^n$ spans $\mathbf{R}^n$, and is therefore a basis for $\mathbf{R}^n$. By the same result, any set of n vectors which spans $\mathbf{R}^n$ must be linearly independent, and therefore is a basis for $\mathbf{R}^n$.

Example 11.3. The sets $\left\{ \begin{bmatrix} 1 \\ 0 \end{bmatrix}, \begin{bmatrix} 0 \\ 1 \end{bmatrix}, \begin{bmatrix} 2 \\ 3 \end{bmatrix} \right\}$ and $\left\{ \begin{bmatrix} 1 \\ 2 \end{bmatrix} \right\}$ are not bases for $\mathbf{R}^2$. The first is a linearly dependent set and the second does not span $\mathbf{R}^2$. $\qquad\diamond$

The proof that every subspace of $\mathbf{R}^n$ has a basis is somewhat technical, and can be found in Appendix B. Since almost every subspace we will encounter is either the null space or column space of a matrix, we demonstrate how to find bases for these subspaces.

Finding a Basis for the Null Space

Given a matrix A we find a spanning set for $N(A)$ by finding $R = \mathrm{rref}(A)$ and solving $R\mathbf{x} = \mathbf{0}$. It turns out that the spanning set found by this method is always linearly independent and thus forms a basis for $N(A)$.

Example 11.4. Let

$$A = \begin{bmatrix} 1 & 1 & 1 & 1 & 1 \\ 1 & 1 & 2 & 4 & 4 \end{bmatrix}.$$

Then

$$R = \operatorname{rref}(A) = \begin{bmatrix} 1 & 1 & 0 & -2 & -2 \\ 0 & 0 & 1 & 3 & 3 \end{bmatrix}.$$

Solving $R\mathbf{x} = \mathbf{0}$ for the pivot variables x_1 and x_3 in terms of the free variables x_2, x_4 and x_5 we get

$$\begin{bmatrix} x_1 \\ x_2 \\ x_3 \\ x_4 \\ x_5 \end{bmatrix} = \begin{bmatrix} -x_2 + 2x_4 + 2x_5 \\ x_2 \\ -3x_4 - 3x_5 \\ x_4 \\ x_5 \end{bmatrix} = x_2 \begin{bmatrix} -1 \\ 1 \\ 0 \\ 0 \\ 0 \end{bmatrix} + x_4 \begin{bmatrix} 2 \\ 0 \\ -3 \\ 1 \\ 0 \end{bmatrix} + x_5 \begin{bmatrix} 2 \\ 0 \\ -3 \\ 0 \\ 1 \end{bmatrix}$$

so

$$N(A) = \operatorname{span}\left(\begin{bmatrix} -1 \\ 1 \\ 0 \\ 0 \\ 0 \end{bmatrix}, \begin{bmatrix} 2 \\ 0 \\ -3 \\ 1 \\ 0 \end{bmatrix}, \begin{bmatrix} 2 \\ 0 \\ -3 \\ 0 \\ 1 \end{bmatrix} \right).$$

It is clear that these vectors are linearly independent since each contains a 1 in an entry which is zero in the other two vectors. These entries correspond to the appearance of the free variables x_2, x_4 and x_5 in entries 2, 4, and 5 of the solution. ◇

In general, if $R = \operatorname{rref}(A)$, the solution of $R\mathbf{x} = \mathbf{0}$ is always a linear combination in which the coefficients are the free variables. The vector $\mathbf{v}_k$ whose coefficient is the free variable x_k has a 1 in the kth component, while each of the remaining vectors will have a zero in the kth component. Thus $\mathbf{v}_k$ cannot be a linear combination of the other vectors. Since this is true for each free variable, the spanning vectors are linearly independent and thus form a basis for the null space of A.

Finding a Basis for the Column Space

By definition, the columns of A span the column space of A. The question is whether or not these vectors are linearly independent and, if not, which vectors should be removed to form a basis. The answer can be determined by looking at the reduced row echelon form R of A. The key is that, although the column spaces of A and R may be different, they share the same null space.

Example 11.5. Let

$$A = \underbrace{\begin{bmatrix} 1 & 0 & -1 & 0 & 4 \\ 2 & 1 & 0 & 0 & 9 \\ -1 & 2 & 5 & 1 & -5 \\ 1 & -1 & -3 & -2 & 9 \end{bmatrix}}_{\mathbf{v}_1 \quad \mathbf{v}_2 \quad \mathbf{v}_3 \quad \mathbf{v}_4 \quad \mathbf{v}_5}.$$

71

The reduced row echelon form of A is

$$R = \begin{bmatrix} 1 & 0 & -1 & 0 & 4 \\ 0 & 1 & 2 & 0 & 1 \\ 0 & 0 & 0 & 1 & -3 \\ 0 & 0 & 0 & 0 & 0 \end{bmatrix}.$$

$$\underbrace{\qquad}_{\mathbf{w}_1 \quad \mathbf{w}_2 \quad \mathbf{w}_3 \quad \mathbf{w}_4 \quad \mathbf{w}_5}$$

It is clear that the columns of R which contain pivots form a linearly independent set, since each such column contains a 1 where the others contain a zero. Furthermore, the remaining columns can be expressed as linear combinations of $\mathbf{w}_1$, $\mathbf{w}_2$ and $\mathbf{w}_4$. In particular, $\mathbf{w}_3 = -1\mathbf{w}_1 + 2\mathbf{w}_2$ and $\mathbf{w}_5 = 4\mathbf{w}_1 + \mathbf{w}_2 - 3\mathbf{w}_4$. So $\{\mathbf{w}_1, \mathbf{w}_2, \mathbf{w}_4\}$ spans the column space of R and is therefore a basis for the column space of R.

These vectors however *not* form a basis for the column space of A. It turns out that the *corresponding columns* of A form a basis for the column space of A. That is,

$$\{\mathbf{v}_1, \mathbf{v}_2, \mathbf{v}_4\} = \left\{ \begin{bmatrix} 1 \\ 2 \\ -1 \\ 1 \end{bmatrix}, \begin{bmatrix} 0 \\ 1 \\ 2 \\ -1 \end{bmatrix}, \begin{bmatrix} 0 \\ 0 \\ 1 \\ -2 \end{bmatrix} \right\}$$

is a basis for the column space of A. To see that $\{\mathbf{v}_1, \mathbf{v}_2, \mathbf{v}_4\}$ spans $C(A)$ we need to express $\mathbf{v}_3$ and $\mathbf{v}_5$ as linear combinations of these vectors. Since the null spaces of R and A are the same, exactly the same relations hold among the columns of R and the columns of A. The fact that $\mathbf{w}_3 = -1\mathbf{w}_1 + 2\mathbf{w}_2$ implies that

$$\begin{bmatrix} -1 \\ 2 \\ -1 \\ 0 \\ 0 \end{bmatrix}$$

is in $N(R) = N(A)$, so $\mathbf{v}_3 = -\mathbf{v}_1 + 2\mathbf{v}_2$ (Check!). Likewise, because $\mathbf{w}_5 = 4\mathbf{w}_1 + \mathbf{w}_2 - 3\mathbf{w}_4$ it follows that $\mathbf{v}_5 = 4\mathbf{v}_1 + \mathbf{v}_2 - 3\mathbf{v}_4$. To show that $\{\mathbf{v}_1, \mathbf{v}_2, \mathbf{v}_4\}$ is linearly independent, suppose that

$$c_1\mathbf{v}_1 + c_2\mathbf{v}_2 + c_4\mathbf{v}_4 = \mathbf{0}.$$

This means that the vector

$$\begin{bmatrix} c_1 \\ c_2 \\ 0 \\ c_4 \\ 0 \end{bmatrix}$$

is in the null space of A, and thus in the null space of R. Hence

$$c_1\mathbf{w}_1 + c_2\mathbf{w}_2 + c_4\mathbf{w}_4 = \mathbf{0}.$$

But since $\{\mathbf{w}_1, \mathbf{w}_2, \mathbf{w}_4\}$ is linearly independent, this implies $c_1 = c_2 = c_4 = 0$. $\diamond$

We now summarize the observations made in this example.

Proposition 11.2. Given any matrix A, the columns of $R = \text{rref}(A)$ which contain pivots form a basis for $C(R)$, and the *corresponding columns* of A form a basis for $C(A)$.

Exercises

In Exercises 1 through 4, find bases for the column space and null space of each of the given matrices.

11.1. $\begin{bmatrix} 1 & 2 & 1 & 0 & 5 \\ 2 & 2 & 0 & 2 & 4 \\ 3 & 2 & 0 & 0 & 3 \\ 4 & 2 & 0 & -2 & 2 \\ 5 & 2 & 0 & -4 & 1 \end{bmatrix}$

11.2. $\begin{bmatrix} 1 & 0 & 2 & 3 & 1 & -1 \\ 2 & -1 & -2 & 5 & 4 & 0 \\ 3 & -1 & 0 & 8 & 5 & -1 \\ 4 & -1 & 2 & 1 & 8 & 0 \end{bmatrix}$

11.3. $\begin{bmatrix} 1 & 2 & 0 & 1 \\ 3 & -1 & -1 & 2 \\ -2 & 5 & 6 & 3 \\ 5 & 5 & 4 & 8 \\ 0 & 2 & 5 & 4 \end{bmatrix}$

11.4. $\begin{bmatrix} 1 & 1 & 1 & 1 & 0 \\ 1 & 1 & 1 & 0 & 1 \\ 1 & 1 & 0 & 1 & 1 \\ 1 & 0 & 1 & 1 & 1 \end{bmatrix}$

In Exercises 5 through 11 find a basis for the given set.

11.5. $\text{span} \left(\begin{bmatrix} 2 \\ 1 \\ 3 \end{bmatrix}, \begin{bmatrix} -4 \\ 3 \\ 8 \end{bmatrix}, \begin{bmatrix} -14 \\ 3 \\ -3 \end{bmatrix} \right)$

11.6. $\text{span} \left(\begin{bmatrix} 1 \\ -5 \\ 7 \\ 4 \end{bmatrix}, \begin{bmatrix} 2 \\ 0 \\ -3 \\ 4 \end{bmatrix}, \begin{bmatrix} 1 \\ 2 \\ 5 \\ -3 \end{bmatrix}, \begin{bmatrix} 1 \\ 3 \\ 0 \\ -2 \end{bmatrix} \right)$

11.7. span $\left(\begin{bmatrix} 2 \\ 1 \\ 0 \\ 1 \\ 0 \end{bmatrix}, \begin{bmatrix} 1 \\ 0 \\ 1 \\ 0 \\ 2 \end{bmatrix}, \begin{bmatrix} 4 \\ 1 \\ 2 \\ 1 \\ 4 \end{bmatrix}, \begin{bmatrix} 0 \\ 1 \\ -2 \\ 1 \\ -4 \end{bmatrix} \right)$

11.8. The plane $3x + 2y - 4z = 0$ in $\mathbf{R}^3$.

11.9. The hyperplane $ax + by + cz + dw = 0$ in $\mathbf{R}^4$.

11.10. The intersection of the planes $2x + 3y + z = 0$ and $3x - 2y + z = 0$.

11.11. The set of solutions of

$$
\begin{array}{rcrcrcrcrcrcl}
x_1 & + & x_2 & + & x_3 & - & x_4 & - & x_5 & - & x_6 & = & 0 \\
2x_1 & + & 2x_2 & - & x_3 & + & x_4 & + & 2x_5 & + & x_6 & = & 0 \\
-x_1 & - & x_2 & + & 2x_3 & - & 2x_4 & - & 3x_5 & - & 2x_6 & = & 0
\end{array}
$$

In Exercises 12 through 15, determine whether the statement is true or false. If true, explain why. If false, give a counterexample.

11.12. Let A be an $m \times n$ matrix and let $R = \mathrm{rref}(A)$. Then if $S = \{\mathbf{v}_1, \ldots, \mathbf{v}_k\}$ is a basis for $C(R)$, then S is a basis for $C(A)$.

11.13. Let A be an $m \times n$ matrix and let $R = \mathrm{rref}(A)$. Then if $S = \{\mathbf{v}_1, \ldots, \mathbf{v}_k\}$ is a basis for $N(R)$, then S is a basis for $N(A)$.

11.14. If $\{\mathbf{v}_1, \mathbf{v}_2, \mathbf{v}_3\}$ is a basis for a subspace V of $\mathbf{R}^n$, then $\{2\mathbf{v}_1 - \mathbf{v}_2, \mathbf{v}_1 + \mathbf{v}_2 + 2\mathbf{v}_3, \mathbf{v}_1 + \mathbf{v}_3\}$ is also a basis for V.

11.15. If $\{\mathbf{v}_1, \mathbf{v}_2, \mathbf{v}_3\}$ is a basis for a subspace V of $\mathbf{R}^n$, then $\{3\mathbf{v}_1 - 2\mathbf{v}_2 + \mathbf{v}_3, -\mathbf{v}_1 + 4\mathbf{v}_2 - \mathbf{v}_3, 2\mathbf{v}_1 + \mathbf{v}_2 + 4\mathbf{v}_3\}$ is also a basis for V.

12 Dimension of a Subspace

The **dimension** of a nontrivial subspace V is the number of elements in any basis for V and is denoted $\dim(V)$. The trivial subspace is defined to have dimension zero.

Example 12.1. The standard basis $\{\mathbf{e}_1, \mathbf{e}_2, \ldots, \mathbf{e}_n\}$ for $\mathbf{R}^n$ has n elements, so the dimension of $\mathbf{R}^n$ is n. $\diamond$

Example 12.2. The plane P defined by $x - 4y + 5z = 0$ has a basis

$$
\left\{ \begin{bmatrix} 4 \\ 1 \\ 0 \end{bmatrix}, \begin{bmatrix} -5 \\ 0 \\ 1 \end{bmatrix} \right\}
$$

so $\dim(P) = 2$.

74

Before we proceed any further there is an important issue that needs to be addressed. A subspace may have many different bases. What if one basis had more elements than another? What would the dimension of the subspace be then? The answer is that this cannot happen. The proof of this relies upon the following fact.

Proposition 12.1. If a set of m vectors spans a subspace V and $n > m$ then any set of n vectors in V must be linearly dependent.

Proof. Suppose

$$V = \text{span}(\mathbf{v}_1, \mathbf{v}_2, \ldots, \mathbf{v}_m)$$

and let $\{\mathbf{w}_1, \mathbf{w}_2, \ldots, \mathbf{w}_n\}$ be any set of vectors in V with $n > m$. We can then write

$$\mathbf{w}_1 = a_{11}\mathbf{v}_1 + a_{21}\mathbf{v}_2 + \cdots + a_{m1}\mathbf{v}_m$$
$$\mathbf{w}_2 = a_{12}\mathbf{v}_1 + a_{22}\mathbf{v}_2 + \cdots + a_{m2}\mathbf{v}_m$$
$$\vdots$$
$$\mathbf{w}_n = a_{1n}\mathbf{v}_1 + a_{2n}\mathbf{v}_2 + \cdots + a_{mn}\mathbf{v}_m$$

for some real numbers a_{ij}. The m by n matrix

$$A = \begin{bmatrix} a_{11} & a_{12} & \cdots & a_{1n} \\ a_{21} & a_{22} & \cdots & a_{2n} \\ \vdots & \vdots & & \vdots \\ a_{m1} & a_{m2} & \cdots & a_{mn} \end{bmatrix}$$

has more columns than rows, so there cannot be a pivot in each column of $\text{rref}(A)$. So by Proposition 8.3, A must have a nontrivial null space. Let

$$\mathbf{c} = \begin{bmatrix} c_1 \\ c_2 \\ \vdots \\ c_n \end{bmatrix}$$

be any nonzero element of $N(A)$. Then

$$c_1\mathbf{w}_1 + c_2\mathbf{w}_2 + \cdots + c_n\mathbf{w}_n = (a_{11}c_1 + a_{12}c_2 + \cdots + a_{1n}c_n)\mathbf{v}_1$$
$$+ (a_{21}c_1 + a_{22}c_2 + \cdots + a_{2n}c_n)\mathbf{v}_2$$
$$\vdots$$
$$+ (a_{m1}c_1 + a_{m2}c_2 + \cdots + a_{mn}c_n)\mathbf{v}_m$$
$$= 0\mathbf{v}_1 + 0\mathbf{v}_2 + \cdots + 0\mathbf{v}_m = \mathbf{0}$$

since the coefficients of $\mathbf{v}_1, \mathbf{v}_2, \ldots, \mathbf{v}_m$ are exactly the components of $A\mathbf{c}$, which equals $\mathbf{0}$ since $\mathbf{c}$ is in the null space of A. Since $\mathbf{c}$ is nonzero, this implies that the set $\{\mathbf{w}_1, \mathbf{w}_2, \ldots, \mathbf{w}_n\}$ is linearly dependent. $\square$

We can now prove that the definition of dimension makes sense.

> **Proposition 12.2.** Let V be a subspace of $\mathbf{R}^n$. Then every basis for V has the same number of elements.

Proof. Suppose $\mathcal{B}_1 = \{\mathbf{v}_1, \ldots, \mathbf{v}_{d_1}\}$ and $\mathcal{B}_2 = \{\mathbf{w}_1, \ldots, \mathbf{w}_{d_2}\}$ are bases for V. Since $\mathcal{B}_1$ spans V and $\mathcal{B}_2$ is linearly independent, it follows from Proposition 12.1 that $d_2 \leq d_1$. On the other hand, since $\mathcal{B}_2$ spans V and $\mathcal{B}_1$ is linearly independent, it follows that $d_1 \leq d_2$. Hence $d_1 = d_2$. $\qquad\square$

Once the dimension d of a subspace V is known, to verify that a set of d elements of V is a basis, it suffices to verify *either* that the set spans V *or* that it is linearly independent. The other property comes for free.

> **Proposition 12.3.** Let V be a subspace with dimension d, and let $S = \{\mathbf{v}_1, \mathbf{v}_2, \ldots, \mathbf{v}_d\}$ be any set of d vectors in V. Then the following statements are equivalent.
>
> 1. S is linearly independent.
>
> 2. S spans V.
>
> 3. S is a basis for V.

That is, any one of the statements implies the other two. Thus, for example, any set of n linearly independent vectors in $\mathbf{R}^n$ is a basis for $\mathbf{R}^n$. Also, any set of n vectors which spans $\mathbf{R}^n$ is a basis for $\mathbf{R}^n$.

Proof. By definition, 3 implies 1 and 2, and 1 and 2 together imply 3. Thus we only need to show that 1 and 2 are equivalent.

First suppose 1 holds, so S is linearly independent. Let $\mathbf{v}$ be any vector in V and consider the set $\{\mathbf{v}, \mathbf{v}_1, \ldots, \mathbf{v}_d\}$. This set consists of $d + 1$ vectors in V, and V can be spanned by d vectors, so by Proposition 12.1 this set must be linearly dependent. Thus

$$c_0 \mathbf{v} + c_1 \mathbf{v}_1 + \cdots + c_d \mathbf{v}_d = \mathbf{0}.$$

Since S is linearly independent, c_0 must be nonzero. Thus

$$\mathbf{v} = -\frac{1}{c_0}(c_1 \mathbf{v}_1 + \cdots + c_d \mathbf{v}_d)$$

so $\mathbf{v}$ is in $\operatorname{span}(\mathbf{v}_1, \ldots, \mathbf{v}_d)$. Since $\mathbf{v}$ is arbitrary, this proves that $V = \operatorname{span}(\mathbf{v}_1, \ldots, \mathbf{v}_d)$. Thus 1 implies 2.

Now suppose 2 holds, and that S is linearly dependent. Then one of the $\mathbf{v}_i$ is a linear combination of the others, so V can be spanned by $d - 1$ vectors. But any basis for V consists of d linearly independent vectors. This contradicts Proposition 12.1, so S must be linearly independent. So 2 implies 1, and thus 1 and 2 are equivalent. $\qquad\square$

We conclude this section by considering the dimensions of the column space and null space of a matrix. Let A be an m by n matrix. We define the **rank** (or column rank) of A by

$$\text{rank}(A) = \dim(C(A)),$$

the dimension of the column space of A, and the **nullity** of A by

$$\text{nullity}(A) = \dim(N(A)),$$

the dimension of the null space of A. We have seen that the columns of A which correspond to the pivot columns of $R = \text{rref}(A)$ form a basis for the column space of A. Thus the rank of A equals the number of pivots in $\text{rref}(A)$. We have also seen that there is one basis vector of the null space of A for each free variable of $A\mathbf{x} = \mathbf{0}$. Thus the nullity of A is simply the number of free variables. Since there are n variables, and the number of pivot variables is $\text{rank}(A)$, the nullity of A must be $n - \text{rank}(A)$. We therefore have the following result.

Proposition 12.4. (Rank-Nullity Theorem) Let A be any m by n matrix. Then

$$\text{rank}(A) + \text{nullity}(A) = n.$$

That is,

$$\dim(C(A)) + \dim(N(A)) = n.$$

Exercises

In Exercises 1 through 11 compute the rank and nullity of the matrix A and state what this implies about the existence and/or uniqueness of solutions $\mathbf{x}$ of $A\mathbf{x} = \mathbf{b}$.

12.1. $A = \begin{bmatrix} 1 & 2 & 3 \\ 1 & 1 & 1 \end{bmatrix}$

12.2. $A = \begin{bmatrix} 1 & 2 \\ 2 & 1 \end{bmatrix}$

12.3. $A = \begin{bmatrix} 1 & 1 \\ 2 & 2 \\ -1 & -1 \end{bmatrix}$

12.4. $A = \begin{bmatrix} 0 & 1 \\ 1 & 0 \\ 2 & 2 \end{bmatrix}$

12.5. $A = \begin{bmatrix} -1 & 2 & 4 \\ 2 & -4 & -8 \end{bmatrix}$

12.6. $A = \begin{bmatrix} 1 & 1 & 1 \\ 1 & 2 & 3 \\ 3 & 2 & 1 \end{bmatrix}$

12.7. $A = \begin{bmatrix} 0 & 1 & 1 & 0 \\ 2 & -1 & -1 & 2 \\ 1 & 2 & 2 & 4 \end{bmatrix}$

12.8. $A = \begin{bmatrix} 7 & 3 & 9 & 1 \\ 3 & 7 & 1 & 9 \\ 9 & 1 & 13 & -3 \\ 1 & 9 & -3 & 13 \end{bmatrix}$

12.9. $A = \begin{bmatrix} 1 & -1 & 2 & -2 \\ -1 & 1 & -2 & 2 \\ 2 & -2 & 4 & -4 \\ -2 & 2 & -4 & 4 \end{bmatrix}$

12.10. $A = \begin{bmatrix} 1 & -2 & 3 & 1 \\ 0 & 0 & 0 & 1 \\ 2 & 1 & 2 & 1 \\ -1 & 0 & -1 & 1 \end{bmatrix}$

12.11. $A = \begin{bmatrix} 1 & -2 & 3 & 1 \\ 0 & 0 & 0 & 1 \\ 2 & 1 & 2 & 1 \\ -1 & -3 & 1 & 1 \end{bmatrix}$

12.12. Evaluate each of the following assertions as always true, sometimes true, or never true.

(a) A set of 3 vectors in $\mathbf{R}^4$ is linearly independent.

(b) A set of 3 vectors in $\mathbf{R}^4$ spans $\mathbf{R}^4$.

(c) A set of 4 vectors in $\mathbf{R}^3$ is linearly independent.

(d) A set of 4 vectors in $\mathbf{R}^3$ spans $\mathbf{R}^3$.

(e) A set of 4 vectors which spans $\mathbf{R}^4$ is linearly independent.

(f) A set of 4 linearly independent vectors in $\mathbf{R}^4$ spans $\mathbf{R}^4$.

12.13. Determine whether each of the following statements is true or false.

(a) For a subspace V with dimension d, any d vectors which span V form a basis for V.

(b) For a subspace V with dimension d, any d vectors which are linearly independent form a basis for V.

(c) If $\{\mathbf{v}_1, \dots, \mathbf{v}_k\}$ span a subspace V, then V is k-dimensional.

(d) If $\{\mathbf{v}_1, \dots, \mathbf{v}_k\}$ is a basis for a subspace V, then any k linear combinations of $\mathbf{v}_1$ through $\mathbf{v}_k$ forms a basis for V.

(e) If $\{\mathbf{v}_1, \dots, \mathbf{v}_k\}$ is a basis for a subspace V, then every basis for V consists of k linear combinations of $\mathbf{v}_1$ through $\mathbf{v}_k$.

13 Linear Transformations

Thus far we have been concerned with the linear subsets of $\mathbf{R}^n$ (subspaces). We now turn our attention to linear functions on $\mathbf{R}^n$. A **function** from a set X to a set Y is a rule f which assigns to each element x of X an element of Y, which we denote by $f(x)$. The set X is called the **domain** of f and the set Y is called the **codomain** of f. The domain can be thought of as the set of all possible inputs for f, while the codomain is the set in which the outputs lie. We write $f : X \to Y$ to denote that f is a function with domain X and codomain Y. We will be mostly concerned with functions whose domain is $\mathbf{R}^n$, or some subset of $\mathbf{R}^n$, and whose codomain is $\mathbf{R}^m$, or a subset of $\mathbf{R}^m$. Functions whose codomain is a subset of $\mathbf{R}$ are called **real valued**, or **scalar valued**. Functions whose codomain is a subset of $\mathbf{R}^m$ for $m \geq 2$ are called **vector valued**, and are denoted using boldface letters.

Example 13.1. The formula $\mathbf{f}(x_1, x_2, x_3) = (x_1^2 + x_2^2 + x_3^2, x_1 x_2 x_3)$ defines a function from $\mathbf{R}^3$ to $\mathbf{R}^2$. $\diamond$

By making the usual identification of points in $\mathbf{R}^n$ with vectors in $\mathbf{R}^n$, we may view a function $\mathbf{f} : \mathbf{R}^n \to \mathbf{R}^m$ as a function which sends vectors in $\mathbf{R}^n$ to vectors in $\mathbf{R}^m$. Linear functions are precisely those functions which are well behaved with respect to the vector operations of addition and scalar multiplication. A **linear transformation** (linear function) is a function $\mathbf{T} : \mathbf{R}^n \to \mathbf{R}^m$ which satisfies

1. $\mathbf{T}(\mathbf{x} + \mathbf{y}) = \mathbf{T}(\mathbf{x}) + \mathbf{T}(\mathbf{y})$

2. $\mathbf{T}(c\mathbf{x}) = c\mathbf{T}(\mathbf{x})$

for all $\mathbf{x}$ and $\mathbf{y}$ in $\mathbf{R}^n$ and all scalars c in $\mathbf{R}$. The first property says that adding $\mathbf{x}$ and $\mathbf{y}$ and then applying $\mathbf{T}$ is equivalent to applying $\mathbf{T}$ to $\mathbf{x}$ and $\mathbf{y}$ and then adding the resulting vectors.

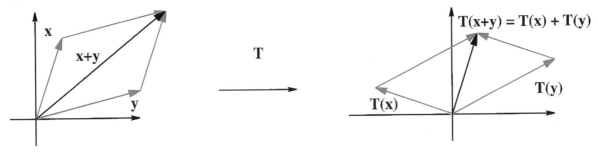

The second property says that scaling first and then applying $\mathbf{T}$ is equivalent to applying $\mathbf{T}$ first and then scaling.

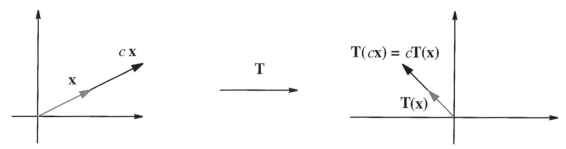

Example 13.2. Let $\mathbf{T} : \mathbf{R}^2 \to \mathbf{R}^2$ be defined by $\mathbf{T}(x_1, x_2) = (2x_1 - 3x_2, x_1 + x_2)$. Since

$$\mathbf{T}(\mathbf{x} + \mathbf{y}) = \begin{bmatrix} 2(x_1 + y_1) - 3(x_2 + y_2) \\ (x_1 + y_1) + (x_2 + y_2) \end{bmatrix} = \begin{bmatrix} 2x_1 - 3x_2 \\ x_1 + x_2 \end{bmatrix} + \begin{bmatrix} 2y_1 - 3y_2 \\ y_1 + y_2 \end{bmatrix} = \mathbf{T}(\mathbf{x}) + \mathbf{T}(\mathbf{y})$$

and

$$\mathbf{T}(c\mathbf{x}) = \begin{bmatrix} 2(cx_1) - 3(cx_2) \\ cx_1 + cx_2 \end{bmatrix} = c \begin{bmatrix} 2x_1 - 3x_2 \\ x_1 + x_2 \end{bmatrix} = c\mathbf{T}(\mathbf{x}),$$

$\mathbf{T}$ is a linear transformation. $\diamond$

Example 13.3. Let $\mathbf{f} : \mathbf{R}^3 \to \mathbf{R}^2$ be the function defined in Example 13.1. Making the arbitrary choice

$$\mathbf{x} = \begin{bmatrix} 1 \\ 1 \\ 1 \end{bmatrix}$$

we see that $\mathbf{f}(\mathbf{x}) = \mathbf{f}(1, 1, 1) = (3, 1)$, while $\mathbf{f}(2\mathbf{x}) = \mathbf{f}(2, 2, 2) = (12, 8)$. Thus $\mathbf{f}(2\mathbf{x}) \neq 2\mathbf{f}(\mathbf{x})$, so $\mathbf{f}$ is not a linear transformation, since the second property fails. $\diamond$

Matrices and Linear Transformations

The properties which define linear transformations are identical to those of matrix-vector products. It follows therefore that multiplication by any matrix defines a linear transformation.

Proposition 13.1. Let A be an $m \times n$ matrix, and let $\mathbf{T}$ be the function from $\mathbf{R}^n$ to $\mathbf{R}^m$ defined by $\mathbf{T}(\mathbf{x}) = A\mathbf{x}$. Then $\mathbf{T}$ is a linear transformation.

Proof. By Proposition 7.1,

$$\mathbf{T}(\mathbf{x} + \mathbf{y}) = A(\mathbf{x} + \mathbf{y}) = A\mathbf{x} + A\mathbf{y} = \mathbf{T}(\mathbf{x}) + \mathbf{T}(\mathbf{y})$$

and

$$\mathbf{T}(c\mathbf{x}) = A(c\mathbf{x}) = cA\mathbf{x} = c\mathbf{T}(\mathbf{x})$$

for any vectors $\mathbf{x}$ and $\mathbf{y}$ in $\mathbf{R}^n$ and any scalar c in $\mathbf{R}$. $\square$

Example 13.4. The linear transformation $\mathbf{T} : \mathbf{R}^3 \to \mathbf{R}^2$ defined by multiplication by the matrix

$$A = \begin{bmatrix} 2 & -3 & 4 \\ 6 & 1 & -2 \end{bmatrix}$$

is given by $\mathbf{T}(x_1, x_2, x_3) = (2x_1 - 3x_2 + 4x_3, 6x_1 + x_2 - 2x_3)$. $\diamond$

The remarkable fact is that every linear transformation is equivalent to multiplication by some matrix.

Proposition 13.2. Let $\mathbf{T} : \mathbf{R}^n \to \mathbf{R}^m$ be a linear transformation. Then $\mathbf{T}$ is equivalent to multiplication by the matrix

$$A = \begin{bmatrix} | & | & & | \\ \mathbf{T}(\mathbf{e}_1) & \mathbf{T}(\mathbf{e}_2) & \cdots & \mathbf{T}(\mathbf{e}_n) \\ | & | & & | \end{bmatrix}$$

where $\mathbf{e}_1$ through $\mathbf{e}_n$ are the standard basis vectors in $\mathbf{R}^n$. That is, $\mathbf{T}(\mathbf{x}) = A\mathbf{x}$ for all $\mathbf{x}$ in $\mathbf{R}^n$.

We say that the **matrix of a linear transformation T** (with respect to the standard basis) is the matrix A defined above.

Proof. First, recall that any vector $\mathbf{x}$ may be written in terms of the standard basis vectors as

$$\mathbf{x} = x_1\mathbf{e}_1 + x_2\mathbf{e}_2 + \cdots + x_n\mathbf{e}_n.$$

Using the linearity of $\mathbf{T}$ we then have

$$\begin{aligned} \mathbf{T}(\mathbf{x}) &= \mathbf{T}(x_1\mathbf{e}_1 + x_2\mathbf{e}_2 + \cdots + x_n\mathbf{e}_n) \\ &= x_1\mathbf{T}(\mathbf{e}_1) + x_2\mathbf{T}(\mathbf{e}_2) + \cdots + x_n\mathbf{T}(\mathbf{e}_n). \end{aligned}$$

The final expression is exactly the product of the matrix A with the vector $\mathbf{x}$. $\square$

Example 13.5. The linear transformation $\mathbf{T} : \mathbf{R}^2 \to \mathbf{R}^2$ defined in Example 13.2 satisfies

$$\mathbf{T}(\mathbf{e}_1) = \begin{bmatrix} 2 \\ 1 \end{bmatrix} \quad \text{and} \quad \mathbf{T}(\mathbf{e}_2) = \begin{bmatrix} -3 \\ 1 \end{bmatrix},$$

so the matrix for $\mathbf{T}$ is

$$A = \begin{bmatrix} 2 & -3 \\ 1 & 1 \end{bmatrix},$$

and thus $\mathbf{T}(\mathbf{x}) = A\mathbf{x}$ for all $\mathbf{x}$ in $\mathbf{R}^2$. $\diamond$

Images and Preimages

Given a function $f : X \to Y$ and a subset S of X, each element of S is then sent by f to some element $f(x)$ of Y. The set of all such elements $f(x)$ is called the **image** of S under f, and is denoted $f(S)$. That is,

$$f(S) = \{f(x) \mid x \text{ is in } S\}.$$

Example 13.6. Let $\mathbf{T}$ be the linear transformation from Example 13.2 and let the line L be parametrized by $\{\mathbf{x}_0 + t\mathbf{v} \mid t \in \mathbf{R}\}$, where

$$\mathbf{x}_0 = \begin{bmatrix} 0 \\ -1 \end{bmatrix} \quad \text{and} \quad \mathbf{v} = \begin{bmatrix} 1 \\ 1 \end{bmatrix}.$$

By linearity

$$\mathbf{T}(\mathbf{x}_0 + t\mathbf{v}) = \mathbf{T}(\mathbf{x}_0) + t\mathbf{T}(\mathbf{v}),$$

so

$$\mathbf{T}(L) = \{\mathbf{T}(\mathbf{x}_0) + t\mathbf{T}(\mathbf{v}) \mid t \in \mathbf{R}\}.$$

This is precisely the line passing through $\mathbf{T}(\mathbf{x}_0)$, parallel to the line spanned by $\mathbf{T}(\mathbf{v})$. Since

$$\mathbf{T}(\mathbf{x}_0) = \begin{bmatrix} 2 & -3 \\ 1 & 1 \end{bmatrix} \begin{bmatrix} 0 \\ -1 \end{bmatrix} = \begin{bmatrix} 3 \\ -1 \end{bmatrix} \quad \text{and} \quad \mathbf{T}(\mathbf{v}) = \begin{bmatrix} 2 & -3 \\ 1 & 1 \end{bmatrix} \begin{bmatrix} 1 \\ 1 \end{bmatrix} = \begin{bmatrix} -1 \\ 2 \end{bmatrix}$$

the image of L under $\mathbf{T}$ is the line which passes through the points $(3, -1)$ and $(2, 1)$.

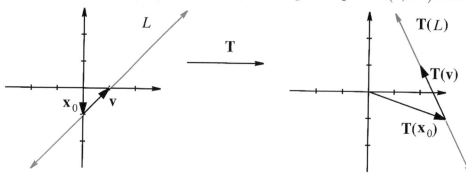

In general, the image under a linear transformation $\mathbf{T}$ of a line $\{\mathbf{x}_0 + t\mathbf{v} \mid t \in \mathbf{R}\}$ is a line $\mathbf{T}(\mathbf{x}_0) + t\mathbf{T}(\mathbf{v})$, unless $\mathbf{T}(\mathbf{v})$ equals the zero vector, in which case the image consists only of the point $\mathbf{T}(\mathbf{x}_0)$. By the same reasoning, it follows that the image under a linear transformation $\mathbf{T}$ of the line segment

$$\{\mathbf{x}_0 + t(\mathbf{x}_1 - \mathbf{x}_0) \mid 0 \le t \le 1\}$$

between $\mathbf{x}_0$ and $\mathbf{x}_1$ is the line segment

$$\{\mathbf{T}(\mathbf{x}_0) + t(\mathbf{T}(\mathbf{x}_1) - \mathbf{T}(\mathbf{x}_0)) \mid 0 \le t \le 1\}$$

between $\mathbf{T}(\mathbf{x}_0)$ and $\mathbf{T}(\mathbf{x}_1)$. Thus the image of any polygonal region may be found by determining the image of its vertices and then "connecting the dots" in order with line segments.

Example 13.7. Consider again the linear transformation $\mathbf{T}$ from Example 13.2, and let R be the region bounded by the polygon with vertices $(0,0)$, $(1,1)$, $(1,-1)$, $(-1,-1)$, and $(-1,1)$, in that order. Since $\mathbf{T}(0,0) = (0,0)$, $\mathbf{T}(1,1) = (-1,2)$, $\mathbf{T}(1,-1) = (5,0)$, $\mathbf{T}(-1,-1) = (1,-2)$ and $\mathbf{T}(-1,1) = (-5,0)$, the image of R under $\mathbf{T}$ is the polygon shown below.

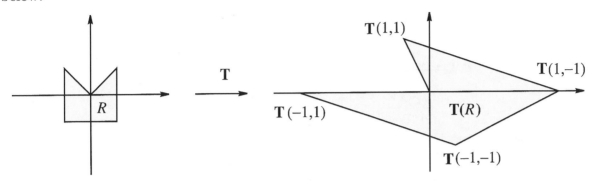

We next show that the image of a subspace under a linear transformation is a subspace.

> **Proposition 13.3.** Let $\mathbf{T} : \mathbf{R}^n \to \mathbf{R}^m$ be a linear transformation, and let V be a subspace of $\mathbf{R}^n$. Then $\mathbf{T}(V)$ is a subspace of $\mathbf{R}^m$.

Proof. We need to show that the image $\mathbf{T}(V)$ is closed under addition and scalar multiplication. So consider two elements $\mathbf{T}(\mathbf{x})$ and $\mathbf{T}(\mathbf{y})$ in $\mathbf{T}(V)$, where $\mathbf{x}$ and $\mathbf{y}$ are necessarily in V. By linearity, their sum is

$$\mathbf{T}(\mathbf{x}) + \mathbf{T}(\mathbf{y}) = \mathbf{T}(\mathbf{x} + \mathbf{y}).$$

Since V is a subspace, $\mathbf{x} + \mathbf{y}$ is in V, so $\mathbf{T}(\mathbf{x} + \mathbf{y})$, and hence $\mathbf{T}(\mathbf{x}) + \mathbf{T}(\mathbf{y})$, is in $\mathbf{T}(V)$. So $\mathbf{T}(V)$ is closed under addition. Next, for any scalar c we have

$$c\mathbf{T}(\mathbf{x}) = \mathbf{T}(c\mathbf{x}).$$

Once again using the fact that V is a subspace, since $\mathbf{x}$ is in V it follows that $c\mathbf{x}$ is in V. Thus $\mathbf{T}(c\mathbf{x})$, and hence $c\mathbf{T}(\mathbf{x})$, is in $\mathbf{T}(V)$, so $\mathbf{T}(V)$ is closed under scalar multiplication. $\square$

We leave it as an exercise to verify that if $\{\mathbf{v}_1, \ldots, \mathbf{v}_k\}$ spans V then $\{\mathbf{T}(\mathbf{v}_1), \ldots, \mathbf{T}(\mathbf{v}_k)\}$ spans $\mathbf{T}(V)$.

Example 13.8. Let $\mathbf{T} : \mathbf{R}^3 \to \mathbf{R}^3$ be the linear transformation whose matrix is

$$A = \begin{bmatrix} 3 & -1 & 1 \\ 2 & 0 & 2 \\ 1 & 1 & 3 \end{bmatrix}$$

and let V be the plane $x_1 + 4x_2 - 6x_3 = 0$. The vectors

$$\mathbf{v}_1 = \begin{bmatrix} -4 \\ 1 \\ 0 \end{bmatrix} \quad \text{and} \quad \mathbf{v}_2 = \begin{bmatrix} 6 \\ 0 \\ 1 \end{bmatrix}$$

form a basis for (and hence span) V. Thus the vectors

$$\mathbf{T}(\mathbf{v}_1) = \begin{bmatrix} -13 \\ -8 \\ -3 \end{bmatrix} \quad \text{and} \quad \mathbf{T}(\mathbf{v}_2) = \begin{bmatrix} 19 \\ 14 \\ 9 \end{bmatrix}$$

span $\mathbf{T}(V)$. Since $\mathbf{T}(\mathbf{v}_1)$ and $\mathbf{T}(\mathbf{v}_2)$ are linearly independent they form a basis for $\mathbf{T}(V)$. It is not in general true however that if $\{\mathbf{v}_1, \ldots, \mathbf{v}_k\}$ is a basis for a subspace V, then $\{\mathbf{T}(\mathbf{v}_1), \ldots, \mathbf{T}(\mathbf{v}_k)\}$ is a basis for $\mathbf{T}(V)$, since these vectors could be linearly dependent. $\diamond$

As a special case of Proposition 13.3, let V be all of $\mathbf{R}^n$. The image of $\mathbf{R}^n$ under a linear transformation $\mathbf{T} : \mathbf{R}^n \to \mathbf{R}^m$ is called the **image** of $\mathbf{T}$, and is denoted $\mathrm{im}(\mathbf{T})$. That is,

$$\mathrm{im}(\mathbf{T}) = \{\mathbf{T}(\mathbf{x}) \mid \mathbf{x} \in \mathbf{R}^n\}.$$

Now let A be the matrix for $\mathbf{T}$. Then $\mathbf{T}(\mathbf{x}) = A\mathbf{x}$ for each $\mathbf{x}$ in $\mathbf{R}^n$, so that

$$\mathrm{im}(\mathbf{T}) = \{A\mathbf{x} \mid \mathbf{x} \in \mathbf{R}^n\} = C(A).$$

Hence the image of $\mathbf{T}$ is precisely the column space of A.

Another important concept is that of a preimage. Given a function $f : X \to Y$, and a subset S of the codomain Y, we might wish to know which elements x in X are sent by f into S. The **preimage** of S under f is the set

$$f^{-1}(S) = \{x \text{ in } X \mid f(x) \text{ is in } S\}.$$

Example 13.9. Let $\mathbf{T} : \mathbf{R}^2 \to \mathbf{R}^2$ be the linear transformation with matrix

$$A = \begin{bmatrix} 1 & 3 \\ 2 & 6 \end{bmatrix}$$

and let $S = \{(0,0), (1,2)\}$. The preimage of S consists of all points in $\mathbf{R}^2$ which are sent by $\mathbf{T}$ to either of these points. Thus we need to solve the systems

$$A\mathbf{x} = \begin{bmatrix} 0 \\ 0 \end{bmatrix} \quad \text{and} \quad A\mathbf{x} = \begin{bmatrix} 1 \\ 2 \end{bmatrix}.$$

The solutions of these systems are

$$\left\{ t \begin{bmatrix} -3 \\ 1 \end{bmatrix} \,\middle|\, t \in \mathbf{R} \right\} \quad \text{and} \quad \left\{ \begin{bmatrix} 1 \\ 0 \end{bmatrix} + t \begin{bmatrix} -3 \\ 1 \end{bmatrix} \,\middle|\, t \in \mathbf{R} \right\},$$

respectively, so the preimage $f^{-1}(S)$ consists of two parallel lines.

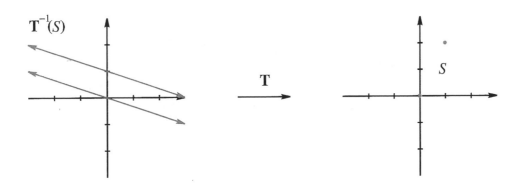

It is left as an exercise to show that the preimage of a subspace under a linear transformation is a subspace. One important example is the preimage of the trivial subspace. The **kernel** of a linear transformation $\mathbf{T} : \mathbf{R}^n \to \mathbf{R}^m$ is the set

$$\ker(\mathbf{T}) = \{\mathbf{x} \text{ in } \mathbf{R}^n \mid \mathbf{T}(\mathbf{x}) = \mathbf{0}\}.$$

If A is the matrix for $\mathbf{T}$, then a vector $\mathbf{x}$ is in the kernel of $\mathbf{T}$ if and only if $A\mathbf{x} = \mathbf{0}$. Hence the kernel of $\mathbf{T}$ is precisely the null space of A.

Sums and Scalar Multiples of Linear Transformations

Given two linear transformations $\mathbf{S} : \mathbf{R}^n \to \mathbf{R}^m$ and $\mathbf{T} : \mathbf{R}^n \to \mathbf{R}^m$, their **sum** is the transformation defined by $(\mathbf{S} + \mathbf{T})(\mathbf{x}) = \mathbf{S}(\mathbf{x}) + \mathbf{T}(\mathbf{x})$. We also define a scalar multiple c of the transformation $\mathbf{T}$ by $(c\mathbf{T})(\mathbf{x}) = c\mathbf{T}(\mathbf{x})$. It is left as an exercise to verify that sums and scalar multiples (and hence linear combinations) of linear transformations are linear transformations. By Proposition 13.2, the jth column of the matrix for $\mathbf{S} + \mathbf{T}$ is

$$(\mathbf{S} + \mathbf{T})(\mathbf{e}_j) = \mathbf{S}(\mathbf{e}_j) + \mathbf{T}(\mathbf{e}_j),$$

the sum of the jth column of the matrix for $\mathbf{S}$ and the jth column of the matrix for $\mathbf{T}$. Since

$$(c\mathbf{S})(\mathbf{e}_j) = c\mathbf{S}(\mathbf{e}_j),$$

for any scalar c, the jth column of the matrix for $c\mathbf{S}$ is c times the jth column of the matrix for $\mathbf{S}$. In view of these facts, we define the sum of two $m \times n$ matrices A and B by

$$A + B = \begin{bmatrix} a_{11} & a_{12} & \cdots & a_{1n} \\ a_{21} & a_{22} & \cdots & a_{2n} \\ \vdots & \vdots & & \vdots \\ a_{m1} & a_{m2} & \cdots & a_{mn} \end{bmatrix} + \begin{bmatrix} b_{11} & b_{12} & \cdots & b_{1n} \\ b_{21} & b_{22} & \cdots & b_{2n} \\ \vdots & \vdots & & \vdots \\ b_{m1} & b_{m2} & \cdots & b_{mn} \end{bmatrix}$$

$$= \begin{bmatrix} a_{11} + b_{11} & a_{12} + b_{12} & \cdots & a_{1n} + b_{1n} \\ a_{21} + b_{21} & a_{22} + b_{22} & \cdots & a_{2n} + b_{2n} \\ \vdots & \vdots & & \vdots \\ a_{m1} + b_{m1} & a_{m2} + b_{m2} & \cdots & a_{mn} + b_{mn} \end{bmatrix}$$

85

and the scalar multiple c of A by

$$cA = c \begin{bmatrix} a_{11} & a_{12} & \cdots & a_{1n} \\ a_{21} & a_{22} & \cdots & a_{2n} \\ \vdots & \vdots & & \vdots \\ a_{m1} & a_{m2} & \cdots & a_{mn} \end{bmatrix} = \begin{bmatrix} ca_{11} & ca_{12} & \cdots & ca_{1n} \\ ca_{21} & ca_{22} & \cdots & ca_{2n} \\ \vdots & \vdots & & \vdots \\ ca_{m1} & ca_{m2} & \cdots & ca_{mn} \end{bmatrix}.$$

We then have the following result.

Proposition 13.4. Suppose $\mathbf{S} : \mathbf{R}^n \to \mathbf{R}^m$ and $\mathbf{T} : \mathbf{R}^n \to \mathbf{R}^m$ are linear transformations with matrices A and B, respectively. Then the matrix for $\mathbf{S} + \mathbf{T}$ is $A + B$ and, for any scalar c, the matrix for $c\mathbf{S}$ is cA.

As with vectors, we denote by $-A$ the matrix $(-1)A$, and by O the **zero matrix**, all of whose entries are zero. With this notation, matrices obey exactly the same properties as vectors.

Proposition 13.5. Let A, B and C be $m \times n$ matrices and let c and d be real numbers. Then

1. $A + B = B + A$
2. $A + (B + C) = (A + B) + C$
3. $A + O = A$
4. $A + (-A) = O$
5. $c(dA) = (cd)A$
6. $(c + d)A = cA + dA$
7. $c(A + B) = cA + cB$
8. $1A = A$

Exercises

In Exercises 1 through 6, for the given function $\mathbf{f} : \mathbf{R}^n \to \mathbf{R}^m$, determine whether the function is a linear transformation. If it is, find the matrix A such that $\mathbf{f}(\mathbf{x}) = A\mathbf{x}$ for all $\mathbf{x} \in \mathbf{R}^n$. If it is not, show that it does not satisfy at least one of the conditions necessary for a function to be a linear transformation.

13.1. $\mathbf{f}(x_1, x_2) = (3x_1, -2x_1 + 5x_2)$

13.2. $\mathbf{f}(x_1, x_2) = (-x_2, x_1 + 1)$

13.3. $\mathbf{f}(x_1, x_2) = (x_1^2 + 3x_2, -5x_1 + 2x_1)$

13.4. $\mathbf{f}(x_1, x_2) = (1, 2)$

13.5. $\mathbf{f}(x_1, x_2) = (0, 0)$

13.6. $\mathbf{f}(x_1, x_2, x_3) = (3x_1 - 2x_3, x_2 + 4x_3, -x_1 + x_2)$

13.7. Suppose $\mathbf{T} : \mathbf{R}^2 \to \mathbf{R}^2$ is a linear transformation satisfying

$$\mathbf{T}\left(\begin{bmatrix} 2 \\ 1 \end{bmatrix}\right) = \begin{bmatrix} 2 \\ 1 \end{bmatrix} \qquad \mathbf{T}\left(\begin{bmatrix} -1 \\ 2 \end{bmatrix}\right) = \begin{bmatrix} 1 \\ -2 \end{bmatrix}.$$

(a) For which matrix A is $\mathbf{T}$ equivalent to multiplication by A?

(b) Describe $\mathbf{T}$ geometrically.

13.8. Suppose $\mathbf{T} : \mathbf{R}^2 \to \mathbf{R}^2$ is a linear transformation satisfying

$$\mathbf{T}\left(\begin{bmatrix} 4 \\ 3 \end{bmatrix}\right) = \begin{bmatrix} 11 \\ 0 \end{bmatrix} \qquad \mathbf{T}\left(\begin{bmatrix} 3 \\ 1 \end{bmatrix}\right) = \begin{bmatrix} 7 \\ -5 \end{bmatrix}.$$

For which matrix A is $\mathbf{T}$ equivalent to multiplication by A?

13.9. Let R_1 be the triangle with vertices $(0,0)$, $(1,1)$ and $(-1,1)$, and let R_2 be the triangle with vertices $(0,0)$, $(2,2)$ and $(2,-2)$. There are exactly two linear transformations $\mathbf{T} : \mathbf{R}^2 \to \mathbf{R}^2$ such that $\mathbf{T}(R_1) = R_2$. Find the matrices for these transformations.

In Exercises 10 through 16 sketch the image of the figure shown below under the linear transformation defined by the given matrix.

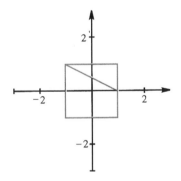

13.10. $\begin{bmatrix} 1 & 0 \\ 0 & 1 \end{bmatrix}$

13.11. $\begin{bmatrix} 1 & 0 \\ 0 & 3 \end{bmatrix}$

13.12. $\begin{bmatrix} 1 & -1 \\ 1 & 1 \end{bmatrix}$

13.13. $\begin{bmatrix} 1 & 2 \\ 2 & 4 \end{bmatrix}$

13.14. $\begin{bmatrix} 0 & 0 \\ 0 & 0 \end{bmatrix}$

13.15. $\begin{bmatrix} 1 & 2 \\ 3 & 4 \end{bmatrix}$

13.16. $\begin{bmatrix} 1 & 2 \\ 0 & 1 \end{bmatrix}$

13.17. Show that if $\{\mathbf{v}_1, \dots, \mathbf{v}_k\}$ spans V then $\{\mathbf{T}(\mathbf{v}_1), \dots, \mathbf{T}(\mathbf{v}_k)\}$ spans $\mathbf{T}(V)$.

13.18. Give an example of a linear transformation $\mathbf{T} : \mathbf{R}^2 \to \mathbf{R}^2$ and a basis $\{\mathbf{v}_1, \mathbf{v}_2\}$ for $\mathbf{R}^2$ such that $\{\mathbf{T}(\mathbf{v}_1), \mathbf{T}(\mathbf{v}_2)\}$ is not a basis for $\mathbf{T}(\mathbf{R}^2)$.

13.19. Given linear transformations $\mathbf{S} : \mathbf{R}^n \to \mathbf{R}^m$ and $\mathbf{T} : \mathbf{R}^n \to \mathbf{R}^m$, show the following.

(a) $\mathbf{S} + \mathbf{T}$ is a linear transformation.

(b) $c\mathbf{S}$ is a linear transformation.

13.20. Show that if $\{\mathbf{v}_1, \mathbf{v}_2, \dots, \mathbf{v}_k\}$ spans V then $\{\mathbf{T}(\mathbf{v}_1), \mathbf{T}(\mathbf{v}_2), \dots, \mathbf{T}(\mathbf{v}_k)\}$ spans $\mathbf{T}(V)$.

13.21. Show by example that it is not always true that if $\{\mathbf{v}_1, \mathbf{v}_2, \dots, \mathbf{v}_k\}$ is a basis for V then $\{\mathbf{T}(\mathbf{v}_1), \mathbf{T}(\mathbf{v}_2), \dots, \mathbf{T}(\mathbf{v}_k)\}$ is a basis for $\mathbf{T}(V)$.

13.22. Show that if $\{\mathbf{v}_1, \mathbf{v}_2, \dots, \mathbf{v}_k\}$ is a linearly dependent set and $\mathbf{T}$ is a linear transformation, then $\{\mathbf{T}(\mathbf{v}_1), \mathbf{T}(\mathbf{v}_2), \dots, \mathbf{T}(\mathbf{v}_k)\}$ is also linearly dependent.

1. Let $\mathbf{T} : \mathbf{R}^n \to \mathbf{R}^m$ be a linear transformation, and let V be a subspace of $\mathbf{R}^m$. Show that the preimage $\mathbf{T}^{-1}(V)$ is a subspace of $\mathbf{R}^n$.

14 Examples of Linear Transformations

In this section we consider a number of important examples of linear transformations. We begin with one of the simplest.

The Identity Transformation

The function $\mathbf{I}_{\mathbf{R}^n} : \mathbf{R}^n \to \mathbf{R}^n$ defined by $\mathbf{I}_{\mathbf{R}^n}(\mathbf{x}) = \mathbf{x}$ is called the **identity transformation** on $\mathbf{R}^n$. It is easy to see that this transformation is linear. Since $\mathbf{I}_{\mathbf{R}^n}(\mathbf{e}_i) = \mathbf{e}_i$ for $i = 1, \dots, n$, the matrix for $\mathbf{I}_{\mathbf{R}^n}$ is

$$\begin{bmatrix} 1 & 0 & \cdots & 0 \\ 0 & 1 & \cdots & 0 \\ \vdots & \vdots & \ddots & \vdots \\ 0 & 0 & \cdots & 1 \end{bmatrix},$$

the $n \times n$ identity matrix I_n. The identity transformation does essentially nothing, sending each vector in $\mathbf{R}^n$ to itself.

Example 14.1. Consider the triangle in $\mathbf{R}^2$ with vertices $(1,1)$, $(-1,1)$ and $(-1,0)$. Its image under the identity transformation is itself.

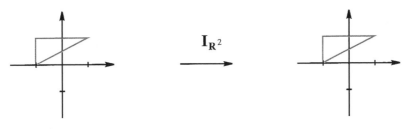

$\diamond$

Scaling Transformations

A **scaling transformation** on $\mathbf{R}^n$ is a transformation $\mathbf{T} : \mathbf{R}^n \to \mathbf{R}^n$ defined by $\mathbf{T}(\mathbf{x}) = \alpha\mathbf{x}$ for some scalar $\alpha \in \mathbf{R}$. Depending on the size of α, scaling transformations either stretch ($|\alpha| > 1$) or contract ($|\alpha| < 1$) vectors, and if α is negative, the transformation reflects vectors through the origin. Since this transformation is α times the identity transformation, its matrix must be

$$\alpha I_n = \begin{bmatrix} \alpha & 0 & \cdots & 0 \\ 0 & \alpha & \cdots & 0 \\ \vdots & \vdots & \ddots & \vdots \\ 0 & 0 & \cdots & \alpha \end{bmatrix}.$$

Example 14.2. The image under $\mathbf{T}$ of the triangle from Example 14.1 is shown below for $\alpha = 2$ and $\alpha = -\frac{1}{2}$.

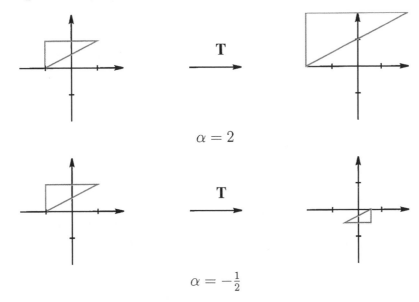

$\diamond$

89

Diagonal Matrices

The matrices of the transformations in the previous examples had zeros in all entries off the main diagonal. Such matrices are called **diagonal** matrices. The most general diagonal matrix takes the form

$$D = \begin{bmatrix} d_1 & 0 & \cdots & 0 \\ 0 & d_2 & \cdots & 0 \\ \vdots & \vdots & \ddots & \vdots \\ 0 & 0 & \cdots & d_n \end{bmatrix}.$$

The transformation $\mathbf{T} : \mathbf{R}^n \to \mathbf{R}^n$ defined by $\mathbf{T}(\mathbf{x}) = D\mathbf{x}$ then has the effect of stretching, contracting or reflecting vectors along the ith coordinate axis by the factor d_i.

Example 14.3. Shown below is the image under $\mathbf{T}$ of the triangle from Example 14.1, with $d_1 = 2$ and $d_2 = -1$.

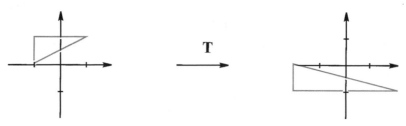

$\Diamond$

Rotations in $\mathbf{R}^2$

Another important type of linear transformation is a **rotation**. We define $\mathbf{Rot}_\theta : \mathbf{R}^2 \to \mathbf{R}^2$ to be the function which rotates vectors in $\mathbf{R}^2$ counterclockwise through the angle θ.

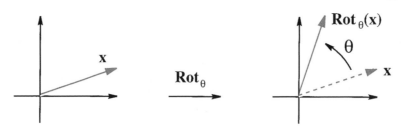

The figure below illustrates that scaling and then rotating is equivalent to rotating and then scaling, so that $\mathbf{Rot}_\theta(c\mathbf{x}) = c\mathbf{Rot}_\theta(\mathbf{x})$.

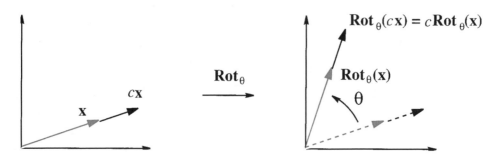

Next notice that adding $\mathbf{x}$ and $\mathbf{y}$ and then rotating is the same as rotating and then adding, so $\mathbf{Rot}_\theta(\mathbf{x} + \mathbf{y}) = \mathbf{Rot}_\theta(\mathbf{x}) + \mathbf{Rot}_\theta(\mathbf{y})$.

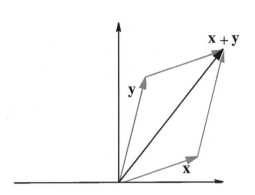

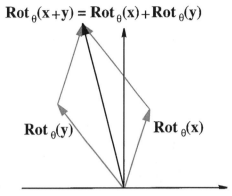

Thus $\mathbf{Rot}_\theta$ is a linear transformation. To find its matrix we evaluate $\mathbf{Rot}_\theta$ on $\mathbf{e}_1$ and $\mathbf{e}_2$ and use Proposition 13.2.

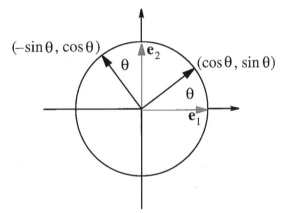

From the figure we see that

$$\mathbf{Rot}_\theta(\mathbf{e}_1) = \begin{bmatrix} \cos\theta \\ \sin\theta \end{bmatrix} \qquad \text{and} \qquad \mathbf{Rot}_\theta(\mathbf{e}_2) = \begin{bmatrix} -\sin\theta \\ \cos\theta \end{bmatrix}.$$

Hence the matrix for $\mathbf{Rot}_\theta$ is

$$\begin{bmatrix} \cos\theta & -\sin\theta \\ \sin\theta & \cos\theta \end{bmatrix}.$$

Example 14.4. The matrix for a $45°$ ($\pi/4$ radian) rotation is

$$\begin{bmatrix} \sqrt{2}/2 & -\sqrt{2}/2 \\ \sqrt{2}/2 & \sqrt{2}/2 \end{bmatrix}.$$

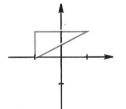

$$\mathbf{Rot}_{\pi/4} \longrightarrow$$

91

Rotations in $\mathbf{R}^3$

Given a line in $\mathbf{R}^3$ which passes through the origin we can define a linear transformation which rotates all points in $\mathbf{R}^3$ about this line through some angle θ. For arbitrarily lines it is quite difficult to determine the matrix for such a transformation. However, if the line happens to be one of the coordinate axes, the calculation is much simpler.

Example 14.5. Let $\mathbf{T} : \mathbf{R}^3 \to \mathbf{R}^3$ be the transformation which rotates vectors in $\mathbf{R}^3$ about the x-axis through angle θ, where the direction is such that the positive y-axis is rotated toward the positive z-axis. Then since the x-axis is fixed by the rotation, $\mathbf{T}(\mathbf{e}_1) = \mathbf{e}_1$. The effect on the yz-plane is that of an ordinary rotation in $\mathbf{R}^2$, so

$$\mathbf{T}(\mathbf{e}_2) = (\cos\theta)\mathbf{e}_2 + (\sin\theta)\mathbf{e}_3 = \begin{bmatrix} 0 \\ \cos\theta \\ \sin\theta \end{bmatrix}$$

$$\mathbf{T}(\mathbf{e}_3) = (-\sin\theta)\mathbf{e}_2 + (\cos\theta)\mathbf{e}_3 = \begin{bmatrix} 0 \\ -\sin\theta \\ \cos\theta \end{bmatrix}.$$

Therefore the matrix for $\mathbf{T}$ is

$$\begin{bmatrix} 1 & 0 & 0 \\ 0 & \cos\theta & -\sin\theta \\ 0 & \sin\theta & \cos\theta \end{bmatrix}.$$

Projections

Given a line L in $\mathbf{R}^n$ which passes through the origin, the **orthogonal projection** of a vector $\mathbf{x}$ onto L is the vector $\mathbf{Proj}_L(\mathbf{x})$ in L such that the difference vector $\mathbf{x} - \mathbf{Proj}_L(\mathbf{x})$ is orthogonal to L.

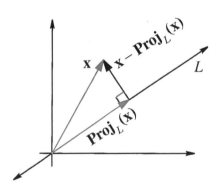

Suppose L is spanned by the vector $\mathbf{v}$. To derive a formula for $\mathbf{Proj}_L(\mathbf{x})$ in terms of $\mathbf{x}$ and $\mathbf{v}$ first observe that $\mathbf{Proj}_L(\mathbf{x}) = c\mathbf{v}$ for some scalar c. Since orthogonal vectors have dot product zero, we have

$$\mathbf{v} \cdot (\mathbf{x} - c\mathbf{v}) = 0.$$

Solving for c we get

$$c = \frac{\mathbf{x} \cdot \mathbf{v}}{\mathbf{v} \cdot \mathbf{v}}$$

and thus we arrive at

$$\mathbf{Proj}_L(\mathbf{x}) = \left(\frac{\mathbf{x} \cdot \mathbf{v}}{\mathbf{v} \cdot \mathbf{v}} \right) \mathbf{v}$$

where $\mathbf{v}$ is any vector which spans the line L. By using a unit vector $\mathbf{u}$ which spans L, the formula simplifies to

$$\mathbf{Proj}_L(\mathbf{x}) = (\mathbf{x} \cdot \mathbf{u})\mathbf{u}.$$

Since

$$
\begin{aligned}
\mathbf{Proj}_L(\mathbf{x} + \mathbf{y}) &= ((\mathbf{x} + \mathbf{y}) \cdot \mathbf{u})\mathbf{u} \\
&= (\mathbf{x} \cdot \mathbf{u} + \mathbf{y} \cdot \mathbf{u})\mathbf{u} \\
&= (\mathbf{x} \cdot \mathbf{u})\mathbf{u} + (\mathbf{y} \cdot \mathbf{u})\mathbf{u} \\
&= \mathbf{Proj}_L(\mathbf{x}) + \mathbf{Proj}_L(\mathbf{y})
\end{aligned}
$$

and

$$
\begin{aligned}
\mathbf{Proj}_L(c\mathbf{x}) &= ((c\mathbf{x}) \cdot \mathbf{u})\mathbf{u} \\
&= c(\mathbf{x} \cdot \mathbf{u})\mathbf{u} \\
&= c\mathbf{Proj}_L(\mathbf{x}),
\end{aligned}
$$

$\mathbf{Proj}_L$ is a linear transformation. We again use Proposition 13.2 to find the matrix for a projection. If L is the line in $\mathbf{R}^2$ spanned by the unit vector

$$\mathbf{u} = \begin{bmatrix} u_1 \\ u_2 \end{bmatrix}$$

then

$$\mathbf{Proj}_L(\mathbf{e}_1) = (\mathbf{e}_1 \cdot \mathbf{u})\mathbf{u} = u_1 \mathbf{u} = \begin{bmatrix} u_1^2 \\ u_1 u_2 \end{bmatrix}$$

$$\mathbf{Proj}_L(\mathbf{e}_2) = (\mathbf{e}_2 \cdot \mathbf{u})\mathbf{u} = u_2 \mathbf{u} = \begin{bmatrix} u_1 u_2 \\ u_2^2 \end{bmatrix}.$$

So the matrix for $\mathbf{Proj}_L$ is

$$\begin{bmatrix} u_1^2 & u_1 u_2 \\ u_1 u_2 & u_2^2 \end{bmatrix}.$$

Example 14.6. If L is the line $y = -x$ then

$$\mathbf{u} = \begin{bmatrix} \sqrt{2}/2 \\ -\sqrt{2}/2 \end{bmatrix}$$

is a unit vector which spans L, so the matrix for $\mathbf{Proj}_L$ is

$$\begin{bmatrix} \frac{1}{2} & -\frac{1}{2} \\ -\frac{1}{2} & \frac{1}{2} \end{bmatrix}.$$

The image under $\mathbf{Proj}_L$ of a triangle is shown below.

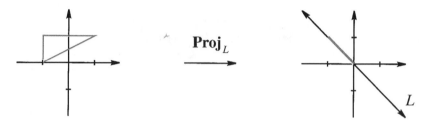

If L is the line in $\mathbf{R}^3$ spanned by the unit vector $\mathbf{u}$ then

$$\mathbf{Proj}_L(\mathbf{e}_1) = \begin{bmatrix} u_1^2 \\ u_1 u_2 \\ u_1 u_3 \end{bmatrix} \qquad \mathbf{Proj}_L(\mathbf{e}_2) = \begin{bmatrix} u_1 u_2 \\ u_2^2 \\ u_2 u_3 \end{bmatrix} \qquad \mathbf{Proj}_L(\mathbf{e}_3) = \begin{bmatrix} u_1 u_3 \\ u_2 u_3 \\ u_3^2 \end{bmatrix}$$

where u_1, u_2 and u_3 are the components of $\mathbf{u}$. Thus the matrix for $\mathbf{Proj}_L$ is

$$\begin{bmatrix} u_1^2 & u_1 u_2 & u_1 u_3 \\ u_1 u_2 & u_2^2 & u_2 u_3 \\ u_1 u_3 & u_2 u_3 & u_3^2 \end{bmatrix}.$$

Reflections

Given a line L in $\mathbf{R}^n$ which passes through the origin, the **reflection** through L of a vector $\mathbf{x}$ is the vector $\mathbf{Ref}_L(\mathbf{x})$ which is the *mirror image* of $\mathbf{x}$ on the opposite side of L.

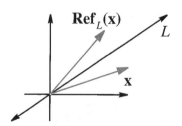

To derive a formula for $\mathbf{Ref}_L$, we let $\mathbf{w} = \mathbf{Proj}_L(\mathbf{x}) - \mathbf{x}$ as shown below.

94

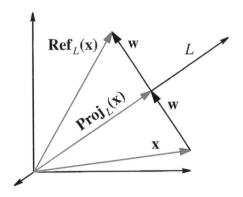

Then

$$\mathbf{Ref}_L(\mathbf{x}) = \mathbf{x} + 2\mathbf{w} = 2\mathbf{Proj}_L(\mathbf{x}) - \mathbf{x} = 2\mathbf{Proj}_L(\mathbf{x}) - \mathbf{I}_{\mathbf{R}^n}(\mathbf{x}) = (2\mathbf{Proj}_L - \mathbf{I}_{\mathbf{R}^n})(\mathbf{x}),$$

so $\mathbf{Ref}_L = 2\mathbf{Proj}_L - \mathbf{I}_{\mathbf{R}^n}$. Since $\mathbf{Proj}_L$ is linear and the identity function is linear, it follows that $\mathbf{Ref}_L$ is a linear transformation, since a linear combination of linear transformations is a linear transformation.

Example 14.7. Let L be the line $y = -x$. Using the result of Example 14.6, it follows that the matrix for $\mathbf{Ref}_L = 2\mathbf{Proj}_L - \mathbf{I}_{\mathbf{R}^2}$ is

$$2\begin{bmatrix} \frac{1}{2} & -\frac{1}{2} \\ -\frac{1}{2} & \frac{1}{2} \end{bmatrix} - \begin{bmatrix} 1 & 0 \\ 0 & 1 \end{bmatrix} = \begin{bmatrix} 0 & -1 \\ -1 & 0 \end{bmatrix}.$$

The image under $\mathbf{Ref}_L$ of a triangle is shown below.

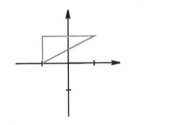 $\quad$ **Ref**$_L$ $\longrightarrow$ $\quad$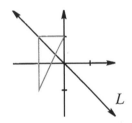

$\diamond$

Other Examples

The linear transformations we have considered thus far are interesting geometrically, but by no means do they represent all possible linear transformations. There are many linear transformations which do not fall into any of the categories listed above. Here is one such example.

Example 14.8. Consider the linear transformation $\mathbf{T} : \mathbf{R}^2 \to \mathbf{R}^2$ whose matrix is

$$A = \begin{bmatrix} 1 & 0 \\ 1 & 1 \end{bmatrix}.$$

This gives the formula $\mathbf{T}(x_1, x_2) = (x_1, x_1 + x_2)$. Thus $\mathbf{T}$ fixes the first component and adds the first component to the second. Points on the (vertical) x_2-axis are therefore left fixed, while points to the right of the axis are shifted upward and points to the left are shifted downward. The image under $\mathbf{T}$ of a triangle is shown below.

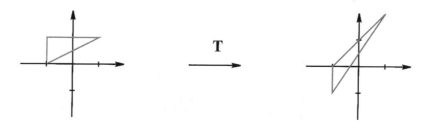

This transformation is an example of a **shear**.   ◇

In the next section we will see how linear transformations can be composed to form new linear transformations.

Exercises

In Exercises 1 through 9, (a) sketch the graph of the figure shown under the linear transformation described and (b) find the matrix A associated with this linear transformation.

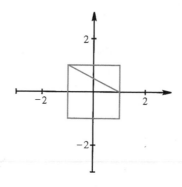

14.1. Counterclockwise rotation by $\pi/3$ radians.

14.2. Rotation by π radians.

14.3. Reflection across the y-axis.

14.4. Reflection across the line $y = \frac{1}{2}x$.

14.5. Projection onto the x-axis.

14.6. Projection onto the line $y = 2x$.

14.7. Scaling by a factor $\alpha = 2$.

14.8. Scaling by a factor $\alpha = 1/2$.

14.9. Reflection through the origin.

14.10. Let $\mathbf{T} : \mathbf{R}^2 \rightarrow \mathbf{R}^2$ be the linear transformation which stretches vectors by a factor 2 along the x-axis and reflects vectors across the x-axis. Find the matrix A associated with this transformation.

14.11. Let $\mathbf{T} : \mathbf{R}^2 \to \mathbf{R}^2$ be the linear transformation which leaves points on the x-axis fixed, but shifts points off the x-axis horizontally by $2y$. Find the matrix A such that $\mathbf{T}$ is equivalent to multiplication by A.

14.12. Let $\mathbf{T} : \mathbf{R}^3 \to \mathbf{R}^3$ be the linear transformation which reflects vectors through the xy-plane. Find the matrix A associated with this transformation.

14.13. Let $\mathbf{Proj}_L : \mathbf{R}^3 \to \mathbf{R}^3$ be the linear transformation which projects vectors onto the line L spanned by $\mathbf{b}$. Find the matrix for $\mathbf{Proj}_L$ for each of the following vectors $\mathbf{b}$.

(a) $\begin{bmatrix} 1 \\ 2 \\ 2 \end{bmatrix}$
(b) $\begin{bmatrix} -3 \\ 0 \\ 4 \end{bmatrix}$
(c) $\begin{bmatrix} 0 \\ 1 \\ 0 \end{bmatrix}$
(d) $\begin{bmatrix} 1 \\ 1 \\ 1 \end{bmatrix}$

14.14. Let $\mathbf{T} : \mathbf{R}^3 \to \mathbf{R}^3$ be the rotation about the y-axis through angle θ, where the direction is such that the positive x-axis is rotated toward the positive z-axis.

14.15. Let $\mathbf{T} : \mathbf{R}^3 \to \mathbf{R}^3$ be the rotation about the z-axis through angle θ, where the direction is such that the positive y-axis is rotated toward the positive x-axis.

15 Composition and Matrix Multiplication

Let $f : X \to Y$ and $g : Y \to Z$ be functions. The **composition** of g with f is the function $g \circ f : X \to Z$ defined by

$$(g \circ f)(x) = g(f(x))$$

for x in X.

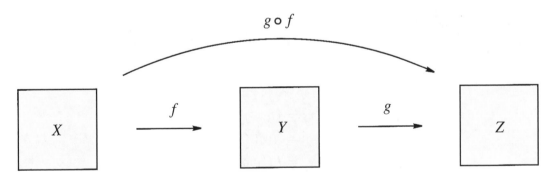

Example 15.1. Let $\mathbf{f} : \mathbf{R}^2 \to \mathbf{R}^2$ and $\mathbf{g} : \mathbf{R}^2 \to \mathbf{R}^2$ be defined by

$$\mathbf{f}(x, y) = (x - y^2, y + x^2)$$
$$\mathbf{g}(x, y) = (y - x, y + x).$$

Then $\mathbf{f} \circ \mathbf{g} : \mathbf{R}^2 \to \mathbf{R}^2$ is given by

$$
\begin{aligned}
(\mathbf{f} \circ \mathbf{g})(x, y) &= \mathbf{f}(\mathbf{g}(x, y)) \\
&= \mathbf{f}(y - x, y + x) \\
&= ((y - x) - (y + x)^2), (y + x) + (y - x)^2) \\
&= (y - x - y^2 - 2xy - x^2, y + x + y^2 - 2xy + x^2).
\end{aligned}
$$

The composition $\mathbf{g} \circ \mathbf{f} : \mathbf{R}^2 \to \mathbf{R}^2$ also makes sense, and is given by

$$
\begin{aligned}
(\mathbf{g} \circ \mathbf{f})(x, y) &= \mathbf{g}(\mathbf{f}(x, y)) \\
&= \mathbf{g}(x - y^2, y + x^2) \\
&= ((y + x^2) - (x - y^2), (y + x^2) + (x - y^2)) \\
&= (y + x^2 - x + y^2, y + x^2 + x - y^2).
\end{aligned}
$$

Notice, however, that $\mathbf{f} \circ \mathbf{g}$ and $\mathbf{g} \circ \mathbf{f}$ are not the same function. $\diamond$

Example 15.2. Let $\mathbf{f} : \mathbf{R} \to \mathbf{R}^3$ be given by $\mathbf{f}(t) = (\cos t, \sin t, t^2)$ and let $\mathbf{g} : \mathbf{R}^3 \to \mathbf{R}^2$ be defined by $\mathbf{g}(x, y, z) = (x^2 + y^2 + z^2, x - y)$. Then $\mathbf{g} \circ \mathbf{f} : \mathbf{R} \to \mathbf{R}^2$ is defined by

$$
\begin{aligned}
(\mathbf{g} \circ \mathbf{f})(t) &= \mathbf{g}(\mathbf{f}(t)) \\
&= \mathbf{g}(\cos t, \sin t, t^2) \\
&= (1 + t^4, \cos t - \sin t),
\end{aligned}
$$

but the composition $\mathbf{f} \circ \mathbf{g}$ is not defined. $\diamond$

These examples illustrate that composition is not in general commutative. Composition is however associative.

> **Proposition 15.1.** Let $f : X \to Y$, $g : Y \to Z$ and $h : Z \to W$. Then
>
> $$(\mathbf{h} \circ \mathbf{g}) \circ \mathbf{f} = \mathbf{h} \circ (\mathbf{g} \circ \mathbf{f}).$$

Proof. Let x be in X. Then

$$
\begin{aligned}
((h \circ g) \circ f)(x) &= (h \circ g)(f(x)) \\
&= h(g(f(x))) \\
&= h((g \circ f)(x)) \\
&= (h \circ (g \circ f))(x).
\end{aligned}
$$

Since this holds for every x in X, the functions $(h \circ g) \circ f$ and $h \circ (g \circ f)$ are the same. $\square$

The parentheses in the formulas above are therefore unnecessary, and we can write without ambiguity $h \circ g \circ f$. For a function $f : X \to X$ it makes sense to compose f with itself. We write f^2 to denote $f \circ f$, f^3 to denote $f \circ f \circ f$, and so on. We now focus our attention on compositions of linear transformations. First we show that any such composition is itself a linear transformation.

Proposition 15.2. Let $\mathbf{T} : \mathbf{R}^k \to \mathbf{R}^n$ and $\mathbf{S} : \mathbf{R}^n \to \mathbf{R}^m$ be linear transformations. Then their composition $\mathbf{S} \circ \mathbf{T} : \mathbf{R}^k \to \mathbf{R}^m$ is also a linear transformation.

Proof. Let $\mathbf{x}$ and $\mathbf{y}$ be in $\mathbf{R}^k$. Then

$$
\begin{aligned}
(\mathbf{S} \circ \mathbf{T})(\mathbf{x} + \mathbf{y}) &= \mathbf{S}(\mathbf{T}(\mathbf{x} + \mathbf{y})) \\
&= \mathbf{S}(\mathbf{T}(\mathbf{x}) + \mathbf{T}(\mathbf{y})) \\
&= \mathbf{S}(\mathbf{T}(\mathbf{x})) + \mathbf{S}(\mathbf{T}(\mathbf{y})) \\
&= (\mathbf{S} \circ \mathbf{T})(\mathbf{x}) + (\mathbf{S} \circ \mathbf{T})(\mathbf{y})
\end{aligned}
$$

and for any scalar c,

$$
\begin{aligned}
(\mathbf{S} \circ \mathbf{T})(c\mathbf{x}) &= \mathbf{S}(\mathbf{T}(c\mathbf{x})) \\
&= \mathbf{S}(c\mathbf{T}(\mathbf{x})) \\
&= c\mathbf{S}(\mathbf{T}(\mathbf{x})) \\
&= c(\mathbf{S} \circ \mathbf{T})(\mathbf{x})
\end{aligned}
$$

so $\mathbf{S} \circ \mathbf{T}$ is linear. $\qquad \square$

Now suppose that $\mathbf{S}$ and $\mathbf{T}$ have matrices A and B, respectively. What is the matrix for $\mathbf{S} \circ \mathbf{T}$ in terms of A and B? The answer is found by evaluating $\mathbf{S} \circ \mathbf{T}$ on the standard basis vectors $\mathbf{e}_1$ through $\mathbf{e}_k$ in $\mathbf{R}^k$. Column j of the matrix for $\mathbf{S} \circ \mathbf{T}$ must be

$$(\mathbf{S} \circ \mathbf{T})(\mathbf{e}_j) = \mathbf{S}(\mathbf{T}(\mathbf{e}_j)) = A(B\mathbf{e}_j).$$

But $B\mathbf{e}_j$ is column j of B, so column j of the matrix for $\mathbf{S} \circ \mathbf{T}$ equals A times column j of B. This motivates the following definition. The **matrix product** AB of an m by n matrix A with an n by k matrix B is the m by k matrix

$$
AB = A \begin{bmatrix} | & | & & | \\ \mathbf{b}_1 & \mathbf{b}_2 & \cdots & \mathbf{b}_k \\ | & | & & | \end{bmatrix} = \begin{bmatrix} | & | & & | \\ A\mathbf{b}_1 & A\mathbf{b}_2 & \cdots & A\mathbf{b}_k \\ | & | & & | \end{bmatrix}
$$

where $\mathbf{b}_1, \mathbf{b}_2, \dots, \mathbf{b}_k$ denote the columns of B. That is, the columns of AB are obtained by multiplying A by the columns of B. Now, since for any vector $\mathbf{x}$, the components of $A\mathbf{x}$ are the dot products of the *rows* of A with $\mathbf{x}$, each entry of AB is the dot product of a row of A with a column of B. More precisely, if $\mathbf{a}_1^T, \mathbf{a}_2^T, \dots, \mathbf{a}_m^T$ denote the rows of A then

$$
AB = \begin{bmatrix} \underline{\quad} & \mathbf{a}_1^T & \underline{\quad} \\ \underline{\quad} & \mathbf{a}_2^T & \underline{\quad} \\ & \vdots & \\ \underline{\quad} & \mathbf{a}_m^T & \underline{\quad} \end{bmatrix} \begin{bmatrix} | & | & & | \\ \mathbf{b}_1 & \mathbf{b}_2 & \cdots & \mathbf{b}_k \\ | & | & & | \end{bmatrix} = \begin{bmatrix} \mathbf{a}_1 \cdot \mathbf{b}_1 & \mathbf{a}_1 \cdot \mathbf{b}_2 & \cdots & \mathbf{a}_1 \cdot \mathbf{b}_k \\ \mathbf{a}_2 \cdot \mathbf{b}_1 & \mathbf{a}_2 \cdot \mathbf{b}_2 & \cdots & \mathbf{a}_2 \cdot \mathbf{b}_k \\ \vdots & \vdots & & \vdots \\ \mathbf{a}_m \cdot \mathbf{b}_1 & \mathbf{a}_m \cdot \mathbf{b}_2 & \cdots & \mathbf{a}_m \cdot \mathbf{b}_k \end{bmatrix}.
$$

Thus the entry in row i, column j of AB is the dot product of row i of A with column j of B. Let a_{ij} and b_{ij} denote the entries in row i column j of A and B, respectively, and let c_{ij} denote the entry in row i column j of the product $C = AB$. Then the ij^{th} entry of the product can be expressed as the sum

$$c_{ij} = \sum_{l=1}^{n} a_{il}b_{lj}.$$

Observe that the product AB is defined only if the number of columns of A equals the number of rows of B. The product has the same number of rows as A and the same number of columns as B. So the product of an m by n matrix with an n by k matrix is an m by k matrix. The heuristic for remembering this is the following:

$$(m \times n)(n \times k) \longrightarrow (m \times k).$$

The adjacent n's cancel.

Example 15.3. Let

$$A = \begin{bmatrix} 2 & 1 & -1 \\ -1 & 3 & 2 \end{bmatrix} \quad \text{and} \quad B = \begin{bmatrix} 1 & 2 & 3 & 4 \\ -1 & 0 & 2 & 1 \\ 3 & 3 & 1 & 2 \end{bmatrix}.$$

Then

$$AB = \begin{bmatrix} -2 & 1 & 7 & 7 \\ 2 & 4 & 5 & 3 \end{bmatrix},$$

while the product BA is not defined since B has 4 columns but A has only 2 rows. ◇

Example 15.4. Let

$$A = \begin{bmatrix} 4 & 1 & -2 \end{bmatrix} \quad \text{and} \quad B = \begin{bmatrix} 1 \\ -1 \\ 3 \end{bmatrix}.$$

Then

$$AB = \begin{bmatrix} -3 \end{bmatrix} \quad \text{and} \quad BA = \begin{bmatrix} 4 & 1 & -2 \\ -4 & -1 & 2 \\ 12 & 3 & -6 \end{bmatrix}.$$

◇

In view of the discussion above, we have the following result.

Proposition 15.3. Let $\mathbf{T} : \mathbf{R}^k \to \mathbf{R}^n$ and $\mathbf{S} : \mathbf{R}^n \to \mathbf{R}^m$ be linear transformations with matrices B and A, respectively. Then the product AB is the matrix for the composition $\mathbf{S} \circ \mathbf{T} : \mathbf{R}^k \to \mathbf{R}^m$.

Example 15.5. Consider the rotations $\mathbf{Rot}_\theta$ and $\mathbf{Rot}_\phi$. The matrix for the composition $\mathbf{Rot}_\theta \circ \mathbf{Rot}_\phi$ is

$$\begin{bmatrix} \cos\theta & -\sin\theta \\ \sin\theta & \cos\theta \end{bmatrix} \begin{bmatrix} \cos\phi & -\sin\phi \\ \sin\phi & \cos\phi \end{bmatrix} = \begin{bmatrix} \cos\theta\cos\phi - \sin\theta\sin\phi & -\cos\theta\sin\phi - \sin\theta\cos\phi \\ \sin\theta\cos\phi + \cos\theta\sin\phi & -\sin\theta\sin\phi + \cos\theta\cos\phi \end{bmatrix}$$
$$= \begin{bmatrix} \cos(\theta+\phi) & -\sin(\theta+\phi) \\ \sin(\theta+\phi) & \cos(\theta+\phi) \end{bmatrix},$$

which, as expected, is the matrix for the rotation through the angle $\theta + \phi$. Thus

$$\mathbf{Rot}_\theta \circ \mathbf{Rot}_\phi = \mathbf{Rot}_{\theta+\phi}.$$

Likewise,

$$\mathbf{Rot}_\phi \circ \mathbf{Rot}_\theta = \mathbf{Rot}_{\theta+\phi}.$$

This is one instance in which composition *does* commute. $\diamond$

Example 15.6. Let L_1 be the x-axis and let L_2 be the line $y = x$. The composition $\mathbf{Proj}_{L_1} \circ \mathbf{Proj}_{L_2}$ has matrix

$$\begin{bmatrix} 1 & 0 \\ 0 & 0 \end{bmatrix} \begin{bmatrix} 1/2 & 1/2 \\ 1/2 & 1/2 \end{bmatrix} = \begin{bmatrix} 1/2 & 1/2 \\ 0 & 0 \end{bmatrix}$$

while the composition $\mathbf{Proj}_{L_2} \circ \mathbf{Proj}_{L_1}$ has matrix

$$\begin{bmatrix} 1/2 & 1/2 \\ 1/2 & 1/2 \end{bmatrix} \begin{bmatrix} 1 & 0 \\ 0 & 0 \end{bmatrix} = \begin{bmatrix} 1/2 & 0 \\ 1/2 & 0 \end{bmatrix}.$$

$\diamond$

We now list some properties of matrix multiplication.

Proposition 15.4. Let A, B, C be matrices for which the sums and products below are defined. Then

1. $A(BC) = (AB)C$.

2. $A(B + C) = AB + AC$.

3. $(A + B)C = AC + BC$.

Proof. The first property follows from the associativity of composition. The others are left as exercises. $\square$

If A is an $n \times n$ matrix, it makes sense to multiply A by itself. In this case we write $A^2 = AA$, $A^3 = AAA$, and so on.

Example 15.7. Let L be the line $y = x$. The matrix for $\mathbf{Proj}_L$ is

$$A = \begin{bmatrix} 1/2 & 1/2 \\ 1/2 & 1/2 \end{bmatrix}.$$

Its square

$$A^2 = \begin{bmatrix} 1/2 & 1/2 \\ 1/2 & 1/2 \end{bmatrix} \begin{bmatrix} 1/2 & 1/2 \\ 1/2 & 1/2 \end{bmatrix} = \begin{bmatrix} 1/2 & 1/2 \\ 1/2 & 1/2 \end{bmatrix} = A$$

is the matrix for $(\mathbf{Proj}_L)^2$. Thus $\mathbf{Proj}_L \circ \mathbf{Proj}_L = \mathbf{Proj}_L$. Think about why this makes sense geometrically. $\diamond$

Exercises

15.1. Compute the following matrix products where possible.

(a) $\begin{bmatrix} 1 & -1 \\ 0 & 1 \end{bmatrix} \begin{bmatrix} 1 & 2 \\ -3 & 4 \end{bmatrix}$ (b) $\begin{bmatrix} 1 & 2 \\ -3 & 4 \end{bmatrix} \begin{bmatrix} 1 & 1 \\ 0 & 1 \end{bmatrix}$ (c) $\begin{bmatrix} 1 & -2 \\ -2 & 4 \end{bmatrix} \begin{bmatrix} 12 & 6 & -2 \\ 6 & 3 & -1 \end{bmatrix}$

(d) $\begin{bmatrix} 1 & 2 & 3 \\ 4 & 5 & 6 \end{bmatrix} \begin{bmatrix} 10 & 11 \\ 12 & 13 \end{bmatrix}$ (e) $\begin{bmatrix} 10 & 11 \\ 12 & 13 \end{bmatrix} \begin{bmatrix} 1 & 2 & 3 \\ 4 & 5 & 6 \end{bmatrix}$ (f) $\begin{bmatrix} 0 & 1 & 0 & 0 \\ 0 & 0 & 1 & 0 \\ 0 & 0 & 0 & 1 \\ 1 & 0 & 0 & 0 \end{bmatrix} \begin{bmatrix} 0 & 0 \\ 9 & 7 \\ 9 & 8 \\ 9 & 9 \end{bmatrix}$

(g) $\begin{bmatrix} 1 & 2 & 3 \end{bmatrix} \begin{bmatrix} 1 \\ 2 \\ 3 \end{bmatrix}$ (h) $\begin{bmatrix} 1 \\ 2 \\ 3 \end{bmatrix} \begin{bmatrix} 1 & 2 & 3 \end{bmatrix}$ (i) $\begin{bmatrix} 1 & 1 \\ 0 & 1 \end{bmatrix}^8$

15.2. For all 2×2 matrices A, B and C, verify directly that $A(BC) = (AB)C$.

1. (a) Prove part (2) of Proposition 15.4.

(b) Prove part (3) of Proposition 15.4.

15.3. Let $\mathbf{T} : \mathbf{R}^2 \to \mathbf{R}^2$ be the linear transformation defined by first rotating counterclockwise through an angle of $\pi/4$ radians and then reflecting across the line $x_2 = -x_1$. What is the matrix for $\mathbf{T}$?

15.4. Let $\mathbf{T} : \mathbf{R}^3 \to \mathbf{R}^3$ be the linear transformation which rotates vectors $30°$ about the x-axis in the direction from the positive y-axis toward the positive z-axis. Let $\mathbf{S} : \mathbf{R}^3 \to \mathbf{R}^3$ be the linear transformation which rotates vectors $45°$ about the z-axis in the direction from the positive x-axis toward the positive y-axis. Find the matrices representing the transformations $\mathbf{T}, \mathbf{S}, \mathbf{T} \circ \mathbf{S}$ and $\mathbf{S} \circ \mathbf{T}$.

15.5. Let $\mathbf{T} : \mathbf{R}^2 \to \mathbf{R}^2$ be rotation in the counterclockwise direction through the angle $\pi/2$ radians followed by projection onto the line which makes an angle θ with the positive x-axis.

(a) Find the matrix for **T** in terms of θ.

(b) What is the kernel of **T**?

(c) What is the image of **T**?

15.6. Let $\mathbf{T} : \mathbf{R}^2 \to \mathbf{R}^2$ be projection onto the line which makes an angle θ with the positive x-axis followed by rotation in the counterclockwise direction through the angle $\pi/2$ radians.

(a) Find the matrix for **T** in terms of θ.

(b) What is the kernel of **T**?

(c) What is the image of **T**?

For Exercises 7 through 10, let $\mathbf{Ref}_\theta : \mathbf{R}^2 \to \mathbf{R}^2$ be reflection across the line which makes an angle θ with the positive x-axis. Let $\mathbf{Rot}_\phi : \mathbf{R}^2 \to \mathbf{R}^2$ be rotation in the counterclockwise direction through the angle ϕ.

15.7. Show that the matrix A associated with $\mathbf{Ref}_\theta$ is given by

$$A = \begin{bmatrix} \cos(2\theta) & \sin(2\theta) \\ \sin(2\theta) & -\cos(2\theta) \end{bmatrix}.$$

15.8. Using the matrix A from Exercise 7,

(a) find the matrix for $\mathbf{Ref}_\theta \circ \mathbf{Rot}_\phi$,

(b) find an angle α (depending on θ and ϕ) such that $\mathbf{Ref}_\theta \circ \mathbf{Rot}_\phi = \mathbf{Ref}_\alpha$.

15.9. Using the matrix A from Exercise 7,

(a) find the matrix for $\mathbf{Rot}_\phi \circ \mathbf{Ref}_\theta$,

(b) find an angle β (depending on θ and ϕ) such that $\mathbf{Rot}_\phi \circ \mathbf{Ref}_\theta = \mathbf{Ref}_\beta$.

15.10. Using the matrix A from Exercise 7,

(a) find the matrix for $\mathbf{Ref}_\theta \circ \mathbf{Ref}_\phi$,

(b) find an angle α (depending on θ and ϕ) such that $\mathbf{Ref}_\theta \circ \mathbf{Ref}_\phi = \mathbf{Rot}_\alpha$.

16 Inverses

The **identity function** on a set X is the function $I_X : X \to X$ such that $I_X(x) = x$ for every x in X. A function $f : X \to Y$ is called **invertible** if there is a function $f^{-1} : Y \to X$ such that

$$f^{-1} \circ f = I_X \quad \text{and} \quad f \circ f^{-1} = I_Y.$$

The function f^{-1}, if it exists, is called the **inverse function** of f.

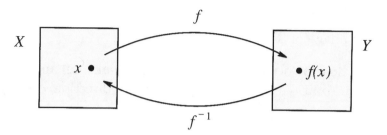

Applying f to x and then applying f^{-1} to the result yields x, so f^{-1} reverses the action of f. Likewise f reverses the action of f^{-1}. The following result allows us to speak of *the* inverse function.

Proposition 16.1. Let $f : X \to Y$ be invertible. Then the inverse of f is unique.

Proof. Suppose that both $g : Y \to X$ and $h : Y \to X$ are both inverses of f. Then,

$$g = I_X \circ g = (h \circ f) \circ g = h \circ (f \circ g) = h \circ I_Y = h.$$

$\square$

Inverse functions are very useful in solving equations of the form $f(x) = y$. The next result states that invertibility is equivalent to the existence and uniqueness of solutions.

Proposition 16.2. A function $f : X \to Y$ is invertible if and only if for every $y \in Y$, the equation $f(x) = y$ has a unique solution x in X.

Proof. First suppose f is invertible. Apply f^{-1} to both sides of $f(x) = y$ to see that

$$x = I_X(x) = (f^{-1} \circ f)(x) = f^{-1}(f(x)) = f^{-1}(y),$$

so $x = f^{-1}(y)$ is the only possible solution of $f(x) = y$. To see that this is in fact a solution, apply f to get

$$f(x) = f(f^{-1}(y)) = (f \circ f^{-1})(y) = I_Y(y) = y.$$

Now suppose that, for every y in Y, the equation $f(x) = y$ has a unique solution x in X. Then define

$$f^{-1}(y) = \text{ the unique solution } x \text{ of } f(x) = y.$$

To verify that this is actually the inverse of f notice that, since $f^{-1}(y)$ is an element x such that $f(x) = y$, we have $f(f^{-1}(y)) = y$. Since this holds for every y in Y, we have $f \circ f^{-1} = I_Y$. Next consider some element x_0 of X. By definition $f^{-1}(f(x_0))$ is the unique solution x of $f(x) = f(x_0)$. But it is clear that the solution must be $x = x_0$, so $f^{-1}(f(x_0)) = x_0$. Since this holds for every x_0 in X, we have $f^{-1} \circ f = I_X$. $\square$

Example 16.1. Let $\mathbf{f} : \mathbf{R}^2 \to \mathbf{R}^2$ be defined by $\mathbf{f}(x_1, x_2) = (x_1 + \sqrt[3]{x_2}, x_1)$. The equation $\mathbf{f}(x_1, x_2) = (y_1, y_2)$ consists of two equations

$$y_1 = x_1 + \sqrt[3]{x_2}$$
$$y_2 = x_1.$$

The second equation implies $x_1 = y_2$. Solving for x_2 in the first equation then gives $x_2 = (y_1 - y_2)^3$. Thus the inverse of $\mathbf{f}$ is given by

$$\mathbf{f}^{-1}(y_1, y_2) = \left(y_2, (y_1 - y_2)^3\right).$$

It is easy now to see that $\mathbf{f} \circ \mathbf{f}^{-1} = \mathbf{f}^{-1} \circ \mathbf{f} = \mathbf{I}_{\mathbf{R}^2}$. $\diamond$

There are two obstacles to invertibility. The first is that the equation $f(x) = y$ may not have solutions for some y.

Example 16.2. Let $\mathbf{f} : \mathbf{R} \to \mathbf{R}^2$ be defined by $\mathbf{f}(x) = (x + 1, 2x - 1)$. Consider the point $(2, 3) \in \mathbf{R}^2$. If we attempt to solve $\mathbf{f}(x) = (2, 3)$ we get the two equations $x + 1 = 2$ and $2x - 1 = 3$. The first implies $x = 1$ while the second implies $x = 2$. Hence there is no x such that $\mathbf{f}(x) = (2, 3)$, and thus $\mathbf{f}$ is not invertible. Geometrically, the image of $\mathbf{R}$ under $\mathbf{f}$ is a line in $\mathbf{R}^2$, so clearly there are many points in $\mathbf{R}^2$ which are not in the range of $\mathbf{f}$. $\diamond$

A function $f : X \to Y$ is **onto** (or **surjective**) if, for every $y \in Y$ there is at least one x in X such that $f(x) = y$. That is, $f(X) = Y$, the image of X under f equals Y. The function in Example 16.2 is not onto. The second obstacle to invertibility is that the equation $f(x) = y$ may have more than one solution for some y.

Example 16.3. Let $\mathbf{f} : \mathbf{R}^3 \to \mathbf{R}^2$ be defined by $\mathbf{f}(x_1, x_2, x_3) = (x_1 + x_2 + x_3, x_1 x_2 x_3)$. Notice that $\mathbf{f}(1, 1, -1) = \mathbf{f}(1, -1, 1) = (1, -1)$, so the equation $\mathbf{f}(\mathbf{x}) = (1, -1)$ has at least two solutions. Therefore $\mathbf{f}$ is not invertible. $\diamond$

A function $f : X \to Y$ is **one to one** (or **injective**) if, for every y in Y there is at most one x in X such that $f(x) = y$. The function in Example 16.3 is not one to one. The following proposition is an immediate consequence of Proposition 16.2.

> **Proposition 16.3.** A function $f : X \to Y$ is invertible if and only if f is both onto and one to one.

Inverse of a Linear Transformation

Next suppose $\mathbf{T} : \mathbf{R}^n \to \mathbf{R}^m$ is a linear transformation. Under what conditions is $\mathbf{T}$ invertible, and if $\mathbf{T}$ is invertible, what is its inverse? To answer the first part of this question, suppose A is the $(m \times n)$ matrix for $\mathbf{T}$. By Proposition 16.3, $\mathbf{T}$ is invertible if and only if $\mathbf{T}$ is both onto and one to one.

For $\mathbf{T}$ to be onto, the equation $\mathbf{T}(\mathbf{x}) = \mathbf{y}$, which is equivalent to the system $A\mathbf{x} = \mathbf{y}$, must have at least one solution $\mathbf{x}$ for each $\mathbf{y}$ in $\mathbf{R}^m$. Recall that the set of $\mathbf{y}$ for which this system has a solution is the column space of A. So $C(A)$ must equal $\mathbf{R}^m$. By Proposition 9.2, this happens if and only if $\text{rref}(A)$ has a pivot in each of its m rows, so $\text{rank}(A) = m$.

Proposition 16.4. The linear transformation $\mathbf{T} : \mathbf{R}^n \to \mathbf{R}^m$ is onto if and only if its matrix A has rank m.

For $\mathbf{T}$ to be one to one, the system $A\mathbf{x} = \mathbf{y}$, must have at most one solution $\mathbf{x}$ for each $\mathbf{y}$ in $\mathbf{R}^m$. By Proposition 8.2 every solution set of $A\mathbf{x} = \mathbf{y}$ is a translation of the null space of A. So in order for solutions to be unique, the null space must be trivial. By Proposition 8.3, this occurs if and only if rref(A) has a pivot in each of its n rows, so rank(A) = n.

Proposition 16.5. The linear transformation $\mathbf{T} : \mathbf{R}^n \to \mathbf{R}^m$ is one to one if and only if its matrix A has rank n.

Example 16.4. Let $\mathbf{T} : \mathbf{R}^2 \to \mathbf{R}^3$ be the linear transformation with matrix

$$A = \begin{bmatrix} 1 & 2 \\ 3 & 4 \\ 5 & 6 \end{bmatrix}.$$

Since

$$\mathrm{rref}(A) = \begin{bmatrix} 1 & 0 \\ 0 & 1 \\ 0 & 0 \end{bmatrix},$$

A has rank 2, so $\mathbf{T}$ is one to one but not onto, and hence not invertible. $\diamond$

Example 16.5. Let $\mathbf{T} : \mathbf{R}^4 \to \mathbf{R}^3$ be the linear transformation with matrix

$$A = \begin{bmatrix} 1 & 1 & 1 & 1 \\ 1 & 2 & 4 & 8 \\ 1 & 3 & 9 & 27 \end{bmatrix}.$$

Since

$$\mathrm{rref}(A) = \begin{bmatrix} 1 & 0 & 0 & 6 \\ 0 & 1 & 0 & -11 \\ 0 & 0 & 1 & 6 \end{bmatrix},$$

A has rank 3, so $\mathbf{T}$ is onto but not one to one, and hence not invertible. $\diamond$

In order for $\mathbf{T}$ to be both onto and one to one, we must have rank(A) = $m = n$, so A must be a square matrix. Thus only linear transformations from $\mathbf{R}^n$ to $\mathbf{R}^n$ can be invertible, and the test for invertibility follows directly from Proposition 9.3.

Proposition 16.6. Let $\mathbf{T} : \mathbf{R}^n \to \mathbf{R}^n$ be a linear transformation with matrix A. Then $\mathbf{T}$ is invertible if and only if rref(A) = I_n.

Example 16.6. Let $\mathbf{T} : \mathbf{R}^3 \to \mathbf{R}^3$ be the linear transformation with matrix

$$A = \begin{bmatrix} 1 & 2 & 3 \\ 4 & 5 & 6 \\ 7 & 8 & 9 \end{bmatrix}.$$

Since

$$\mathrm{rref}(A) = \begin{bmatrix} 1 & 0 & -1 \\ 0 & 1 & 2 \\ 0 & 0 & 0 \end{bmatrix},$$

A has rank 2, and therefore $\mathbf{T}$ is not invertible. $\diamond$

Example 16.7. Let

$$A = \begin{bmatrix} 1 & -1 & -1 \\ -1 & 2 & 3 \\ 1 & 1 & 4 \end{bmatrix}.$$

Since $\mathrm{rref}(A) = I_3$, the linear transformation $\mathbf{T} : \mathbf{R}^3 \to \mathbf{R}^3$ with matrix A is invertible. $\diamond$

Now suppose $\mathbf{T} : \mathbf{R}^n \to \mathbf{R}^n$ is an invertible linear transformation. We leave it as an exercise to verify that its inverse $\mathbf{T}^{-1} : \mathbf{R}^n \to \mathbf{R}^n$ is also a linear transformation. If A is the matrix for $\mathbf{T}$, then we denote by A^{-1} the matrix for $\mathbf{T}^{-1}$. The matrix A^{-1} is called the **inverse matrix** of A, and we say that A is an **invertible matrix**. Since

$$\mathbf{T} \circ \mathbf{T}^{-1} = \mathbf{T}^{-1} \circ \mathbf{T} = \mathbf{I}_{\mathbf{R}^n},$$

and since the matrix for $\mathbf{I}_{\mathbf{R}^n}$ is the $n \times n$ identity matrix I_n, it follows that

$$\boxed{AA^{-1} = A^{-1}A = I_n.}$$

Example 16.8. Geometrically it is clear that the rotation $\mathbf{Rot}_\theta$ is invertible, with inverse $\mathbf{Rot}_{-\theta}$. So

$$A = \begin{bmatrix} \cos\theta & -\sin\theta \\ \sin\theta & \cos\theta \end{bmatrix} \implies A^{-1} = \begin{bmatrix} \cos(-\theta) & -\sin(-\theta) \\ \sin(-\theta) & \cos(-\theta) \end{bmatrix} = \begin{bmatrix} \cos\theta & \sin\theta \\ -\sin\theta & \cos\theta \end{bmatrix}.$$

It is easy to verify directly that $AA^{-1} = A^{-1}A = I_2$. $\diamond$

Next we consider the task of finding the inverse of a linear transformation. Suppose $\mathbf{T} : \mathbf{R}^n \to \mathbf{R}^n$ is an invertible linear transformation. The columns of the matrix for $\mathbf{T}^{-1}$ are

$$\mathbf{x}_1 = \mathbf{T}^{-1}(\mathbf{e}_1) \qquad \mathbf{x}_2 = \mathbf{T}^{-1}(\mathbf{e}_2) \qquad \dots \qquad \mathbf{x}_n = \mathbf{T}^{-1}(\mathbf{e}_n).$$

Thus $\mathbf{x}_1$ through $\mathbf{x}_n$ satisfy

$$\mathbf{T}(\mathbf{x}_1) = \mathbf{e}_1 \qquad \mathbf{T}(\mathbf{x}_2) = \mathbf{e}_2 \qquad \dots \qquad \mathbf{T}(\mathbf{x}_n) = \mathbf{e}_n$$

or, in terms of the matrix A for $\mathbf{T}$,

$$A\mathbf{x}_1 = \mathbf{e}_1 \qquad A\mathbf{x}_2 = \mathbf{e}_2 \qquad \dots \qquad A\mathbf{x}_n = \mathbf{e}_n.$$

The invertibility of $\mathbf{T}$ guarantees a unique solution to each of these equations.

Example 16.9. Let

$$A = \begin{bmatrix} 1 & -1 & -1 \\ -1 & 2 & 3 \\ 1 & 1 & 4 \end{bmatrix}.$$

Solving $A\mathbf{x}_1 = \mathbf{e}_1$ we obtain

$$\begin{bmatrix} 1 & -1 & -1 & | & 1 \\ -1 & 2 & 3 & | & 0 \\ 1 & 1 & 4 & | & 0 \end{bmatrix} \longrightarrow \begin{bmatrix} 1 & 0 & 0 & | & 5 \\ 0 & 1 & 0 & | & 7 \\ 0 & 0 & 1 & | & -3 \end{bmatrix}$$

so

$$\mathbf{x}_1 = \begin{bmatrix} 5 \\ 7 \\ -3 \end{bmatrix}$$

is the first column of A^{-1}. We could continue in the same way to find $\mathbf{x}_2$ and $\mathbf{x}_3$, but it is more efficient to do all the calculations simultaneously. Since the same elimination steps take place in all three calculations, we may simply augment the matrix A with three columns $\mathbf{e}_1$, $\mathbf{e}_2$ and $\mathbf{e}_3$. Doing so we obtain

$$\begin{bmatrix} 1 & -1 & -1 & | & 1 & 0 & 0 \\ -1 & 2 & 3 & | & 0 & 1 & 0 \\ 1 & 1 & 4 & | & 0 & 0 & 1 \end{bmatrix} \longrightarrow \begin{bmatrix} 1 & 0 & 0 & | & 5 & 3 & -1 \\ 0 & 1 & 0 & | & 7 & 5 & -2 \\ 0 & 0 & 1 & | & -3 & -2 & 1 \end{bmatrix}$$

To the right of the divider appear $\mathbf{x}_1$, $\mathbf{x}_2$ and $\mathbf{x}_3$. Thus

$$A^{-1} = \begin{bmatrix} 5 & 3 & -1 \\ 7 & 5 & -2 \\ -3 & -2 & 1 \end{bmatrix}.$$

Observe that $AA^{-1} = A^{-1}A = I_3$. $\diamond$

This example illustrates a general procedure for finding the inverse of a matrix A.

Proposition 16.7. If A is an invertible matrix, then $\mathrm{rref}[A \mid I_n] = [I_n \mid A^{-1}]$.

Example 16.10. Let

$$A = \begin{bmatrix} 1 & 1 & 1 \\ 1 & 2 & 4 \\ 1 & 3 & 9 \end{bmatrix}.$$

Then

$$\begin{bmatrix} 1 & 1 & 1 & | & 1 & 0 & 0 \\ 1 & 2 & 4 & | & 0 & 1 & 0 \\ 1 & 3 & 9 & | & 0 & 0 & 1 \end{bmatrix} \longrightarrow \begin{bmatrix} 1 & 0 & 0 & | & 3 & -3 & 1 \\ 0 & 1 & 0 & | & -5/2 & 4 & -3/2 \\ 0 & 0 & 1 & | & 1/2 & -1 & 1/2 \end{bmatrix},$$

so

$$A^{-1} = \begin{bmatrix} 3 & -3 & 1 \\ -5/2 & 4 & -3/2 \\ 1/2 & -1 & 1/2 \end{bmatrix}.$$

$\diamondsuit$

Inverse functions satisfy the following properties.

Proposition 16.8. Let $f : X \to Y$ and $g : Y \to Z$ be invertible functions. Then

1. f^{-1} is invertible and $(f^{-1})^{-1} = f$,

2. $g \circ f$ is invertible and $(g \circ f)^{-1} = f^{-1} \circ g^{-1}$.

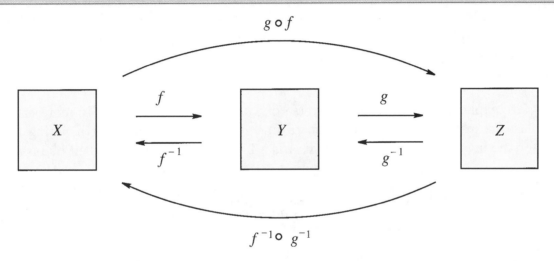

Proof. By definition, f^{-1} is invertible if there exists a function $h : X \to Y$ such that

$$h \circ f^{-1} = I_Y$$
$$f^{-1} \circ h = I_X.$$

But f is exactly such a function, so it is the inverse of f^{-1}. Next, since

$$\begin{aligned} (g \circ f) \circ (f^{-1} \circ g^{-1}) &= g \circ (f \circ f^{-1}) \circ g^{-1} \\ &= g \circ I_Y \circ g^{-1} \\ &= g \circ g^{-1} \\ &= I_Z \end{aligned}$$

and

$$\begin{aligned} (f^{-1} \circ g^{-1}) \circ (g \circ f) &= f^{-1} \circ (g^{-1} \circ g) \circ f \\ &= f^{-1} \circ I_Y \circ f \\ &= f^{-1} \circ f \\ &= I_X \end{aligned}$$

$f^{-1} \circ g^{-1}$ is the inverse of $g \circ f$. $\square$

As an immediate corollary, we have the corresponding results for invertible matrices.

> **Proposition 16.9.** Let A and B be $n \times n$ invertible matrices. Then
>
> 1. A^{-1} is invertible and $(A^{-1})^{-1} = A$,
>
> 2. AB is invertible and $(AB)^{-1} = B^{-1}A^{-1}$.

Repeated application of the second identity yields formulas for inverses of longer products. For instance, $(ABC)^{-1} = ((AB)C)^{-1} = C^{-1}(AB)^{-1} = C^{-1}B^{-1}A^{-1}$.

Example 16.11. Express the inverse of $B^{-1}ACA^{-1}$ as a product of A, B, C and their inverses.

Solution.

$$(B^{-1}ACA^{-1})^{-1} = (A^{-1})^{-1}C^{-1}A^{-1}(B^{-1})^{-1} = AC^{-1}A^{-1}B.$$

$\diamond$

There are explicit formulas for the inverse of a matrix, but it is actually more efficient in most cases to use Gaussian elimination to find inverses. One case worth noting is the formula for the inverse of a 2×2 matrix. If A is a 2×2 invertible matrix then its inverse is

$$A^{-1} = \begin{bmatrix} a & b \\ c & d \end{bmatrix}^{-1} = \frac{1}{ad - bc} \begin{bmatrix} d & -b \\ -c & a \end{bmatrix}.$$

This formula can be verified directly by multiplying by A.

Example 16.12. Let

$$A = \begin{bmatrix} 1 & 2 \\ 3 & 4 \end{bmatrix}.$$

Then

$$A^{-1} = \frac{1}{-2} \begin{bmatrix} 4 & -2 \\ -3 & 1 \end{bmatrix}.$$

$\diamond$

Notice that this formula makes sense only when $ad - bc \neq 0$. The expression $ad - bc$ is the subject of the next section.

Exercises

In Exercises 1 through 10, find the inverses of the given matrix, or demonstrate that the matrix is not invertible.

16.1. $\begin{bmatrix} 1 & 0 & 3 \\ 4 & 4 & 2 \\ 2 & 5 & -4 \end{bmatrix}$

16.2. $\begin{bmatrix} 0 & 1 & 1 \\ 1 & 0 & 1 \\ 1 & 1 & 0 \end{bmatrix}$

16.3. $\begin{bmatrix} 0 & -1 \\ 1 & 0 \end{bmatrix}$

16.4. $\begin{bmatrix} 3 & -1 \\ -3 & 1 \end{bmatrix}$

16.5. $\begin{bmatrix} 3 & -1 \\ 1 & -3 \end{bmatrix}$

16.6. $\begin{bmatrix} 2 & 0 & -1 \\ 1 & 1 & 0 \\ 1 & 1 & 2 \end{bmatrix}$

16.7. $\begin{bmatrix} 1 & 1 & 1 \\ 1 & 2 & 3 \\ 1 & 4 & 9 \end{bmatrix}$

16.8. $\begin{bmatrix} 4 & 2 & 0 \\ 1 & 2 & 1 \\ 1 & -1 & 8 \end{bmatrix}$

16.9. $\begin{bmatrix} 0 & 0 & 3 \\ 0 & -2 & 0 \\ 4 & 0 & 0 \end{bmatrix}$

16.10. $\begin{bmatrix} 3 & 10 & 3 & 8 \\ 3 & -2 & 8 & 7 \\ 2 & 1 & 4 & -5 \\ 5 & 11 & 7 & 3 \end{bmatrix}$

16.11. Simplify the expression $(I_n + A)(I_n - A)$. (Here I_n refers to the $n \times n$ identity matrix, and A is an arbitrary matrix of the same size.)

16.12. Simplify the expression $(B + A)(B - A)$, where A, B are $n \times n$ matrices. Beware!

16.13. Let

$$A = \begin{bmatrix} a & b \\ b & -a \end{bmatrix}$$

and suppose you know that $A^{-1} = A$. What can you say about a and b?

111

16.14. For which choices of the constant k is the matrix

$$\begin{bmatrix} 1 & 1 & 1 \\ 1 & 2 & k \\ 1 & 4 & k^2 \end{bmatrix}$$

invertible?

16.15. Find two invertible matrices whose sum is not invertible.

16.16. Assume A is an invertible matrix.

   (a) If $Ax = \lambda x$ where $\lambda \neq 0$ is a scalar, what is $A^{-1}x$?

   (b) If $Au = v$, what is $A^{-1}(2v)$?

   (c) If $Au_1 = 2v_1$ and $Au_2 = 3v_2$, what is $A^{-1}(3v_1 + 2v_2)$?

16.17. Suppose A is a matrix which satisfies

$$A \begin{bmatrix} 1 \\ 2 \\ 0 \end{bmatrix} = \begin{bmatrix} 2 \\ 4 \\ 0 \end{bmatrix} \qquad A \begin{bmatrix} 1 \\ 0 \\ 1 \end{bmatrix} = \begin{bmatrix} 1 \\ 0 \\ 1 \end{bmatrix} \qquad A \begin{bmatrix} 0 \\ 3 \\ -1 \end{bmatrix} = \begin{bmatrix} 0 \\ -3 \\ 1 \end{bmatrix}$$

and let

$$C = \begin{bmatrix} 1 & 1 & 0 \\ 2 & 0 & 3 \\ 0 & 1 & -1 \end{bmatrix}.$$

   (a) Explain without calculating C^{-1} why

$$C^{-1}AC = \begin{bmatrix} 2 & 0 & 0 \\ 0 & 1 & 0 \\ 0 & 0 & -1 \end{bmatrix}$$

   Hint: The jth column of a matrix M is Me_j, where e_j is the jth standard basis vector. Apply this to the matrix $M = C^{-1}AC$.

   (b) Calculate C^{-1}.

   (c) Using the results from parts (a) and (b), calculate A.

16.18. Suppose $\mathbf{T} : \mathbf{R}^n \to \mathbf{R}^n$ is an invertible linear transformation. Show that $\mathbf{T}^{-1} : \mathbf{R}^n \to \mathbf{R}^n$ is also a linear transformation.

16.19. Consider the linear transformations defined by the following matrices. Which are one to one? Which are onto? For those which are both one to one and onto, find the inverse.

   (a) $\begin{bmatrix} 1 & 2 & 3 \\ 4 & 5 & 6 \end{bmatrix}$ (b) $\begin{bmatrix} 1 & 4 \\ 2 & 5 \\ 3 & 6 \end{bmatrix}$ (c) $\begin{bmatrix} 1 & 2 & 3 \\ 4 & 5 & 6 \\ 7 & 8 & 9 \end{bmatrix}$ (d) $\begin{bmatrix} 1 & 2 & 0 \\ 2 & 3 & 2 \\ 0 & 2 & 1 \end{bmatrix}$

16.20. Which of the following are true for all linear transformations $\mathbf{T} : \mathbf{R}^n \to \mathbf{R}^m$?

(a) If $n < m$ then $\mathbf{T}$ must be onto.

(b) If $n < m$ then $\mathbf{T}$ must be one to one.

(c) If $n < m$ then $\mathbf{T}$ cannot be onto.

(d) If $n < m$ then $\mathbf{T}$ cannot be one to one.

(e) If $n > m$ then $\mathbf{T}$ must be onto.

(f) If $n > m$ then $\mathbf{T}$ must be one to one.

(g) If $n > m$ then $\mathbf{T}$ cannot be onto.

(h) If $n > m$ then $\mathbf{T}$ cannot be one to one.

16.21. Suppose $\mathbf{T} : \mathbf{R}^n \to \mathbf{R}^m$ is a linear transformation that is one to one.

(a) Show that if $\mathbf{T}(\mathbf{v}) = \mathbf{0}$, then $\mathbf{v} = \mathbf{0}$.

(b) Show that if the vectors $\{\mathbf{v}_1, \mathbf{v}_2, \ldots, \mathbf{v}_k\}$ are linearly independent, then so are the vectors $\{\mathbf{T}(\mathbf{v}_1), \mathbf{T}(\mathbf{v}_2), \ldots, \mathbf{T}(\mathbf{v}_k)\}$.

(c) Find an example of a linear transformation $\mathbf{S}$ and linearly independent vectors $\mathbf{v}_1$ and $\mathbf{v}_2$ such that $\mathbf{S}(\mathbf{v}_1)$ and $S(\mathbf{v}_2)$ are linearly dependent.

16.22. Suppose C and A are $n \times n$ matrices and C is invertible.

(a) If $m \geq 1$ explain why $C^{-1} A^m C = (C^{-1}AC)^m$.

(b) If $B = C^{-1}AC$ explain why $CBC^{-1} = A$.

16.23. Let

$$A = \begin{bmatrix} 1 & -2 \\ 1 & 4 \end{bmatrix} \qquad C = \begin{bmatrix} 1 & 2 \\ -1 & -1 \end{bmatrix}.$$

(a) Find C^{-1}.

(b) Calculate $C^{-1}AC$.

(c) Calculate A^7 using the result from part (b). Hint: Let $D = (C^{-1}AC)^7 = C^{-1}A^7C$. Then $CDC^{-1} = A^7$.

17 Determinants

The **determinant** of a 2×2 matrix

$$A = \begin{bmatrix} a & b \\ c & d \end{bmatrix}$$

is the number

$$\det(A) = \begin{vmatrix} a & b \\ c & d \end{vmatrix} = ad - bc.$$

The vertical lines in place of brackets signify the determinant. Notice that the determinant appears in the formula for the inverse of A,

$$A^{-1} = \frac{1}{\det(A)} \begin{bmatrix} d & -b \\ -c & a \end{bmatrix}.$$

Thus if $\det(A)$ is nonzero then A is invertible. It is left as an exercise to verify that A is not invertible if $\det(A) = 0$.

Proposition 17.1. A 2×2 matrix A is invertible if and only if $\det(A) \neq 0$.

Example 17.1. Let

$$A = \begin{bmatrix} 1 & 2 \\ -2 & -4 \end{bmatrix} \qquad \text{and} \qquad B = \begin{bmatrix} 2 & 3 \\ 4 & 0 \end{bmatrix}.$$

Then $\det(A) = 0$ and $\det(B) = -12$, so B is invertible, but A is not invertible. $\diamond$

The determinant of a 3 by 3 matrix is defined as follows.

$$\begin{vmatrix} a_{11} & a_{12} & a_{13} \\ a_{21} & a_{22} & a_{23} \\ a_{31} & a_{32} & a_{33} \end{vmatrix} = a_{11} \begin{vmatrix} a_{22} & a_{23} \\ a_{32} & a_{33} \end{vmatrix} - a_{12} \begin{vmatrix} a_{21} & a_{23} \\ a_{31} & a_{33} \end{vmatrix} + a_{13} \begin{vmatrix} a_{21} & a_{22} \\ a_{31} & a_{32} \end{vmatrix}$$

Each term is an entry from the first row multiplied by the determinant of the 2 by 2 matrix obtained by removing the row and column of that entry from the original 3 by 3 matrix.

Example 17.2.

$$\begin{vmatrix} 1 & 2 & 4 \\ 2 & -1 & 3 \\ 4 & 0 & -1 \end{vmatrix} = 1 \begin{vmatrix} -1 & 3 \\ 0 & -1 \end{vmatrix} - 2 \begin{vmatrix} 2 & 3 \\ 4 & -1 \end{vmatrix} + 4 \begin{vmatrix} 2 & -1 \\ 4 & 0 \end{vmatrix}$$

$$= 1(1) - 2(-14) + 4(4) = 45.$$

$\diamond$

We define recursively the determinant of any $n \times n$ matrix A. Let a_{ij} denote the entry in the ith row, j^{th} column of A, and let A_{ij} denote the $(n-1) \times (n-1)$ matrix obtained by removing the ith row and j^{th} column from A. Then we define

$$\det(A) = a_{11} \det(A_{11}) - a_{12} \det(A_{12}) + \cdots + (-1)^{n+1} a_{1n} \det(A_{1n}).$$

Each term is again an entry from the first row multiplied by the determinant of an $(n-1) \times (n-1)$ matrix, and the signs alternate $+, -, +, -$.

Example 17.3.

$$\begin{vmatrix} 1 & 2 & 3 & 4 \\ 1 & 0 & 2 & 0 \\ 0 & 1 & 2 & 3 \\ 2 & 3 & 0 & 0 \end{vmatrix} = 1 \cdot \begin{vmatrix} 0 & 2 & 0 \\ 1 & 2 & 3 \\ 3 & 0 & 0 \end{vmatrix} - 2 \cdot \begin{vmatrix} 1 & 2 & 0 \\ 0 & 2 & 3 \\ 2 & 0 & 0 \end{vmatrix} + 3 \cdot \begin{vmatrix} 1 & 0 & 0 \\ 0 & 1 & 3 \\ 2 & 3 & 0 \end{vmatrix} - 4 \cdot \begin{vmatrix} 1 & 0 & 2 \\ 0 & 1 & 2 \\ 2 & 3 & 0 \end{vmatrix}$$

$$= 1 \left(0 \cdot \begin{vmatrix} 2 & 3 \\ 0 & 0 \end{vmatrix} - 2 \cdot \begin{vmatrix} 1 & 3 \\ 3 & 0 \end{vmatrix} + 0 \cdot \begin{vmatrix} 1 & 2 \\ 3 & 0 \end{vmatrix} \right) - 2 \left(1 \cdot \begin{vmatrix} 2 & 3 \\ 0 & 0 \end{vmatrix} - 2 \begin{vmatrix} 0 & 3 \\ 2 & 0 \end{vmatrix} + 0 \cdot \begin{vmatrix} 0 & 2 \\ 2 & 0 \end{vmatrix} \right)$$

$$+ 3 \left(1 \cdot \begin{vmatrix} 1 & 3 \\ 3 & 0 \end{vmatrix} - 0 \cdot \begin{vmatrix} 0 & 3 \\ 2 & 0 \end{vmatrix} + 0 \cdot \begin{vmatrix} 0 & 1 \\ 2 & 3 \end{vmatrix} \right) - 4 \left(1 \begin{vmatrix} 1 & 2 \\ 3 & 0 \end{vmatrix} - 0 \cdot \begin{vmatrix} 0 & 2 \\ 2 & 0 \end{vmatrix} + 2 \begin{vmatrix} 0 & 1 \\ 2 & 3 \end{vmatrix} \right)$$

$$= 1(18) - 2(12) + 3(-9) - 4(-10)$$

$$= 7$$

$\Diamond$

In this example the calculations were simplified by the presence of zeros in the first rows of the three by three determinants. It turns out that determinants may be expanded along any row. The formula for expansion along row i is

$$\det(A) = \sum_{j=1}^{n}(-1)^{i+j}a_{ij}\det(A_{ij}).$$

The sign of each term is determined by the following checkerboard pattern.

$$\begin{bmatrix} + & - & + & \cdots \\ - & + & - & \cdots \\ + & - & + & \cdots \\ \vdots & \vdots & \vdots & \ddots \end{bmatrix}$$

Expanding along a row which contains zeros simplifies the calculations.

Example 17.4. Expanding along the fourth row gives

$$\begin{vmatrix} 1 & 2 & 3 & 4 \\ 1 & 0 & 2 & 0 \\ 0 & 1 & 2 & 3 \\ 2 & 3 & 0 & 0 \end{vmatrix} = -2 \cdot \begin{vmatrix} 2 & 3 & 4 \\ 0 & 2 & 0 \\ 1 & 2 & 3 \end{vmatrix} + 3 \cdot \begin{vmatrix} 1 & 3 & 4 \\ 1 & 2 & 0 \\ 0 & 2 & 3 \end{vmatrix},$$

and expanding each of the remaining determinants along the second row gives

$$-2 \cdot 2 \cdot \begin{vmatrix} 2 & 4 \\ 1 & 3 \end{vmatrix} + 3 \left(-1 \cdot \begin{vmatrix} 3 & 4 \\ 2 & 3 \end{vmatrix} + 2 \begin{vmatrix} 1 & 4 \\ 0 & 3 \end{vmatrix} \right) = -2 \cdot 2 \cdot 2 + 3 \cdot (-1 + 6) = 7.$$

$\Diamond$

One may also expand along any column. The formula for expansion along column j is

$$\det(A) = \sum_{i=1}^{n}(-1)^{i+j}a_{ij}\det(A_{ij}).$$

Example 17.5. Expanding the matrix

$$A = \begin{bmatrix} 2 & -1 & 3 & 5 \\ 0 & 3 & 5 & 7 \\ 0 & 0 & -2 & 1 \\ 0 & 0 & 0 & -1 \end{bmatrix}$$

along the first column we get

$$\det(A) = 2 \cdot \begin{vmatrix} 3 & 5 & 7 \\ 0 & -2 & 1 \\ 0 & 0 & -1 \end{vmatrix} = (2) \cdot (3) \cdot \begin{vmatrix} -2 & 1 \\ 0 & -1 \end{vmatrix} = (2) \cdot (3) \cdot (-2) \cdot (-1) = 12,$$

the product of the diagonal entries. $\diamond$

A matrix is called **upper triangular** if all entries below the diagonal are zero. Likewise a matrix is called **lower triangular** if all entries above the diagonal are zero. The determinant of any upper or lower triangular matrix is the product of its diagonal entries. We now list the basic properties of determinants.

Proposition 17.2.

1. (Determinants are Alternating) Exchanging two rows reverses the sign of the determinant. That is, if A and B agree in all but two rows, which are swapped, then $\det(B) = -\det(A)$.

2. (Determinants are Multilinear) The determinant is linear in each row.

   (a) If row i of A is the sum of row i of B and row i of C, and all the other rows of A, B and C agree, then $\det(A) = \det(B) + \det(C)$.

   (b) If row i of A equals a scalar multiple c of row i of B, and all the other rows of A and B agree, then $\det(A) = c \det(B)$.

The first property applies in the case of a single row exchange, with all other rows held fixed. Likewise, the second property applies *one row at a time*, with all other rows held fixed. In particular it does *not* imply that $\det(A+B) = \det(A)+\det(B)$. Determinants are *not* linear, but rather *linear in each row*. One important consequence of the first property is that the determinant of any matrix with two identical rows is necessarily zero.

Example 17.6. Let

$$A = \begin{bmatrix} 1 & 2 & 4 \\ 4 & 5 & -1 \\ 1 & 2 & 4 \end{bmatrix}.$$

116

By Property 1, if we exchange the first and third rows of A, this reverses the sign of the determinant. But since the first and third rows of A are the same, this implies

$$\det(A) = \det \begin{bmatrix} 1 & 2 & 4 \\ 4 & 5 & -1 \\ 1 & 2 & 4 \end{bmatrix} = -\det \begin{bmatrix} 1 & 2 & 4 \\ 4 & 5 & -1 \\ 1 & 2 & 4 \end{bmatrix} = -\det(A)$$

and thus $\det(A) = 0$. $\diamond$

Now suppose that for some matrix A we form a matrix B whose entries are the same as those of A, except in row i, which equals row i of A minus a scalar multiple of row j of A ($j \neq i$). By Property 2,

$$\det(B) = \begin{vmatrix} \vdots \\ \mathbf{v}_i^T - c\mathbf{v}_j^T \\ \vdots \\ \mathbf{v}_j^T \\ \vdots \end{vmatrix} = \begin{vmatrix} \vdots \\ \mathbf{v}_i^T \\ \vdots \\ \mathbf{v}_j^T \\ \vdots \end{vmatrix} - c \begin{vmatrix} \vdots \\ \mathbf{v}_j^T \\ \vdots \\ \mathbf{v}_j^T \\ \vdots \end{vmatrix}.$$

The last matrix has a repeated row, so its determinant is zero, so we are left with

$$\begin{vmatrix} \vdots \\ \mathbf{v}_i^T \\ \vdots \\ \mathbf{v}_j^T \\ \vdots \end{vmatrix} = \det(A).$$

Thus subtracting a multiple of one row from another does not change the determinant. This makes it possible to compute determinants by performing Gaussian elimination. For larger matrices this is actually much more efficient than using the recursive formula.

Example 17.7.

$$\begin{vmatrix} 1 & 2 & 2 & 1 \\ 1 & 2 & 4 & 2 \\ 2 & 7 & 5 & 2 \\ -1 & 4 & -6 & 3 \end{vmatrix} = \begin{vmatrix} 1 & 2 & 2 & 1 \\ 0 & 0 & 2 & 1 \\ 0 & 3 & 1 & 0 \\ 0 & 6 & -4 & 4 \end{vmatrix} = -\begin{vmatrix} 1 & 2 & 2 & 1 \\ 0 & 3 & 1 & 0 \\ 0 & 0 & 2 & 1 \\ 0 & 6 & -4 & 4 \end{vmatrix} = -\begin{vmatrix} 1 & 2 & 2 & 1 \\ 0 & 3 & 1 & 0 \\ 0 & 0 & 2 & 1 \\ 0 & 0 & -6 & 4 \end{vmatrix}$$

$$= -\begin{vmatrix} 1 & 2 & 2 & 1 \\ 0 & 3 & 1 & 0 \\ 0 & 0 & 2 & 1 \\ 0 & 0 & 0 & 7 \end{vmatrix} = -(1)(3)(2)(7) = -42$$

$\diamond$

This method of calculating the determinant also reveals an important connection between the determinant and invertibility. Given an $n \times n$ matrix A, consider the effect on the determinant of each elimination step in going from A to $\mathrm{rref}(A)$. Dividing a row by a nonzero scalar divides the determinant by that scalar, subtracting a multiple of one row from another does not change the determinant, and exchanging two rows reverses the sign of the determinant. At the end, we reach the determinant of $\mathrm{rref}(A)$, so the determinant of A is some nonzero multiple of the determinant of $\mathrm{rref}(A)$. Now recall that A is invertible if and only if $\mathrm{rref}(A) = I_n$. So if A is invertible, then since $\det(I_n) = 1$, it follows that $\det(A)$ is nonzero. On the other hand, if A is not invertible, then $\mathrm{rref}(A)$ has fewer than n pivots, and thus has a row of zeros. Expanding the determinant along that row gives $\det(\mathrm{rref}(A)) = 0$, and consequently $\det(A) = 0$. We therefore have the following generalization of Proposition 17.1.

Proposition 17.3. An $n \times n$ matrix A is invertible if and only if $\det(A) \neq 0$.

The determinant may therefore be used as a test for invertibility.

Example 17.8. The upper triangular matrix

$$A = \begin{bmatrix} 4 & 8 & 2 & -5 \\ 0 & 7 & -2 & 9 \\ 0 & 0 & -3 & 1 \\ 0 & 0 & 0 & 1 \end{bmatrix}$$

has determinant $(4)(7)(-3)(1) = -84$ and is therefore invertible, while in the matrix

$$B = \begin{bmatrix} 3 & 2 & 1 & -2 \\ 1 & -1 & 8 & 7 \\ 6 & 4 & 2 & -4 \\ 2 & 2 & -1 & 4 \end{bmatrix}$$

the third row is twice the first row. Using the linearity of the determinant in the third row gives

$$\det(B) = 2 \begin{vmatrix} 3 & 2 & 1 & -2 \\ 1 & -1 & 8 & 7 \\ 3 & 2 & 1 & -2 \\ 2 & 2 & -1 & 4 \end{vmatrix} = 0$$

since the first and third rows of the new matrix are equal. Thus B is not invertible. $\diamond$

Although determinants are not linear, they are multiplicative.

Proposition 17.4. If A and B are $n \times n$ matrices, then $\det(AB) = \det(A)\det(B)$.

From this we deduce the formula for the determinant of the inverse of a matrix. Suppose A is an $n \times n$ invertible matrix. Then $AA^{-1} = I_n$. Taking the determinant of both sides gives

$$1 = \det(I_n) = \det(AA^{-1}) = \det(A)\det(A^{-1}),$$

so

$$\det(A^{-1}) = \frac{1}{\det(A)}.$$

Determinants and Area

Let $\mathbf{T} : \mathbf{R}^2 \to \mathbf{R}^2$ be the linear transformation whose matrix is

$$A = \begin{bmatrix} a & b \\ c & d \end{bmatrix}$$

and let S denote the square generated by the standard bases vectors $\mathbf{e}_1$ and $\mathbf{e}_2$. The image $\mathbf{T}(S)$ of S under $\mathbf{T}$ is the parallelogram generated by the vectors

$$\mathbf{v}_1 = \mathbf{T}(\mathbf{e}_1) = \begin{bmatrix} a \\ c \end{bmatrix} \quad \text{and} \quad \mathbf{v}_2 = \mathbf{T}(\mathbf{e}_2) = \begin{bmatrix} b \\ d \end{bmatrix}.$$

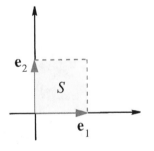
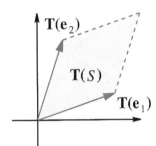

The area of the square S is 1. What is the area of $\mathbf{T}(S)$? The area of a parallelogram is its base times its height, as shown below.

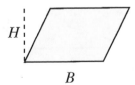
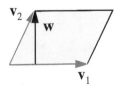

In terms of $\mathbf{v}_1$ and $\mathbf{v}_2$, the base is $B = \|\mathbf{v}_1\|$ and the height is $H = \|\mathbf{w}\|$ where $\mathbf{w} = \mathbf{v}_2 - \mathbf{Proj}_L(\mathbf{v}_2)$ and L is the line spanned by $\mathbf{v}_1$. By the Pythagorean Theorem,

$$\|\mathbf{w}\|^2 = \|\mathbf{v}_2\|^2 - \|\mathbf{Proj}_L(\mathbf{v}_2)\|^2 = \mathbf{v}_2 \cdot \mathbf{v}_2 - \left(\frac{\mathbf{v}_2 \cdot \mathbf{v}_1}{\mathbf{v}_1 \cdot \mathbf{v}_1}\right)^2 \mathbf{v}_1 \cdot \mathbf{v}_1.$$

It is easier to work with B^2H^2, the square of the area. Doing so, we find that

$$B^2H^2 = (\mathbf{v}_1 \cdot \mathbf{v}_1)\left[\mathbf{v}_2 \cdot \mathbf{v}_2 - \left(\frac{\mathbf{v}_2 \cdot \mathbf{v}_1}{\mathbf{v}_1 \cdot \mathbf{v}_1}\right)^2 \mathbf{v}_1 \cdot \mathbf{v}_1\right]$$
$$= (\mathbf{v}_1 \cdot \mathbf{v}_1)(\mathbf{v}_2 \cdot \mathbf{v}_2) - (\mathbf{v}_2 \cdot \mathbf{v}_1)^2$$
$$= (a^2 + c^2)(b^2 + d^2) - (ab + cd)^2$$
$$= a^2d^2 + c^2b^2 - 2abcd$$
$$= (ad - bc)^2$$
$$= (\det(A))^2.$$

Taking square roots we find that $BH = |\det(A)|$.

> **Proposition 17.5.** Let A be any 2 by 2 matrix. The area of the parallelogram generated by the columns of A is $|\det(A)|$.

Example 17.9. Let

$$\mathbf{v} = \begin{bmatrix} -3 \\ 4 \end{bmatrix} \qquad \text{and} \qquad \mathbf{w} = \begin{bmatrix} 5 \\ 2 \end{bmatrix}.$$

The area of the parallelogram formed by $\mathbf{v}$ and $\mathbf{w}$ is

$$\left|\det\begin{bmatrix} -3 & 5 \\ 4 & 2 \end{bmatrix}\right| = |-6 - 20| = 26.$$

$\diamond$

Next, consider the effect of $\mathbf{T}$ on rectangles of different sizes. Let R be the rectangle generated by $c_1\mathbf{e}_1$ and $c_2\mathbf{e}_2$ for some positive scalars c_1 and c_2. The area of R is then c_1c_2, and $\mathbf{T}(R)$ is the parallelogram generated by $c_1\mathbf{T}(\mathbf{e}_1) = c_1\mathbf{v}_1$ and $c_2\mathbf{T}(\mathbf{e}_2) = c_2\mathbf{v}_2$.

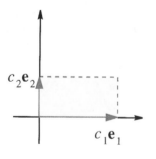
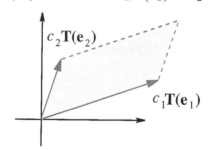

The area of $\mathbf{T}(R)$ is therefore

$$\left|\det\begin{bmatrix} c_1a & c_2b \\ c_1c & c_2d \end{bmatrix}\right| = c_1c_2|\det(A)|,$$

exactly $|\det(A)|$ times the area of R. Thus the determinant of A can be interpreted as an area expansion factor for the transformation $\mathbf{T}$. The remarkable fact is that this property holds for arbitrary regions, not just rectangles.

Proposition 17.6. Let $\mathbf{T}: \mathbf{R}^2 \to \mathbf{R}^2$ be a linear transformation with matrix A, and let R be a region in $\mathbf{R}^2$. Then the area of $\mathbf{T}(R)$ is $|\det(A)|$ times the area of R.

We will not provide a rigorous proof of this fact, but here is the idea of why it is true. Imagine subdividing R into many small rectangles. Then the image of R is subdivided into many small parallelograms, each of which has area $|\det(A)|$ times the area of the corresponding rectangle in R.

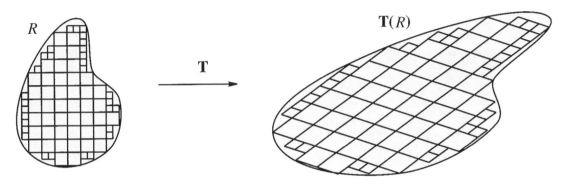

The formula then follows since the area of R is the sum of the areas of these rectangles and the area of $\mathbf{T}(R)$ is the sum of the areas of the parallelograms.

Example 17.10. Let R be the region enclosed by the ellipse

$$\frac{x^2}{4} + y^2 = 1$$

and let $\mathbf{T}$ be the linear transformation defined by the matrix

$$A = \begin{bmatrix} 1 & -1 \\ 1 & 1 \end{bmatrix}.$$

The image of R under $\mathbf{T}$ is also an ellipse. The area of an ellipse with major axis a and minor axis b is πab. Thus the area of R is 2π ($a = 2$ and $b = 1$). Since $|\det A| = 2$, the area of $\mathbf{T}(R)$ is 4π.

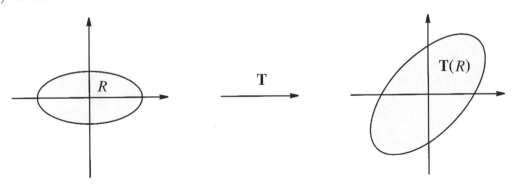

Exercises

Compute the determinant of each matrix in Exercises 1 through 8 and state whether or not the matrix is invertible.

17.1. $\begin{bmatrix} 4 & 6 \\ 4 & 9 \end{bmatrix}$

17.2. $\begin{bmatrix} 4 & 6 \\ 6 & 9 \end{bmatrix}$

17.3. $\begin{bmatrix} 0 & 2 & -1 \\ 3 & 2 & 2 \\ -5 & 1 & 2 \end{bmatrix}$

17.4. $\begin{bmatrix} 1 & 2 & 5 \\ 1 & 2 & 3 \\ 1 & 2 & 1 \end{bmatrix}$

17.5. $\begin{bmatrix} 0 & 0 & 1 \\ 0 & 1 & 0 \\ 1 & 0 & 0 \end{bmatrix}$

17.6. $\begin{bmatrix} 1 & 1 & 2 & 1 \\ 0 & 1 & 3 & 2 \\ 2 & -1 & 0 & 4 \\ 1 & 1 & 1 & 1 \end{bmatrix}$

17.7. $\begin{bmatrix} 1 & 1 & 1 & 1 \\ 2 & 1 & 1 & 2 \\ 0 & 1 & 2 & 4 \\ 3 & 3 & 4 & 5 \end{bmatrix}$

17.8. $\begin{bmatrix} 0 & 1 & 2 & 3 \\ 1 & 2 & 4 & 7 \\ 2 & 4 & 8 & 15 \\ 3 & 7 & 15 & 30 \end{bmatrix}$

17.9. (a) Verify Property 1 of Proposition 17.2 for all 2×2 matrices.

(b) Verify Property 2 of Proposition 17.2 for all 2×2 matrices.

17.10. For all 2×2 matrices A and B, verify that $\det(AB) = \det(A)\det(B)$.

17.11. Find 2×2 matrices A and B such that $\det(A+B) \neq \det(A) + \det(B)$. Find 2×2 matrices A and B such that $\det(A+B) = \det(A) + \det(B)$.

17.12. Let

$$A = \begin{bmatrix} a & b \\ c & d \end{bmatrix}$$

and suppose $ad - bc = 0$. Show that A is not invertible.

17.13. The **trace** of a square matrix A is the sum of the entries on the main diagonal. That is,

$$\operatorname{tr}(A) = \sum_{i=1}^{n} a_{ii}$$

where a_{ii} is the entry in row i column i of A.

(a) Show that for any 2×2 matrix A the following identity holds.

$$A^2 - \operatorname{tr}(A)A + \det(A)I_2 = \begin{bmatrix} 0 & 0 \\ 0 & 0 \end{bmatrix}$$

where I_2 is the 2×2 identity matrix and for scalars c the matrix cA is obtained by multiplying each entry of A by c.

(b) Show for any two $n \times n$ matrices A and B that $\operatorname{tr}(AB) = \operatorname{tr}(BA)$.

17.14. Let A be an $n \times n$ matrix and c any real number. How are $\det(A)$ and $\det(cA)$ related?

17.15. (a) Suppose R is the quadrilateral with vertices $(0,0)$, $(2,1)$, $(3,-3)$ and $(4,-1)$. Find the area of R.

(b) If $\mathbf{T}$ is the linear transformation with matrix

$$A = \begin{bmatrix} -3 & 2 \\ -1 & 2 \end{bmatrix}$$

and R is the region from part (a), find the area of the region $\mathbf{T}(R)$.

17.16. Let $\mathbf{T}$ be the linear transformation given by multiplication by

$$A = \begin{bmatrix} 2 & 1 \\ -1 & 2 \end{bmatrix}$$

and let R be the region bounded by the triangle with vertices $(1,1)$, $(4,-1)$ and $(5,5)$.

(a) Find the area of R.

(b) Find the area of the region $\mathbf{T}(R)$.

17.17. Let A be a 3×3 matrix with columns $\mathbf{u}$, $\mathbf{v}$ and $\mathbf{w}$ in order from left to right. Show that $\det(A) = \mathbf{u} \cdot (\mathbf{v} \times \mathbf{w})$.

17.18. Let C be the cube in $\mathbf{R}^3$ generated by the standard basis vectors $\mathbf{e}_1$, $\mathbf{e}_2$ and $\mathbf{e}_3$, and let $\mathbf{T} : \mathbf{R}^3 \to \mathbf{R}^3$ be a linear transformation with matrix A. Use the result of the previous exercise to show that the volume of $\mathbf{T}(C)$ is $|\det(A)|$. See Exercise 26 of Section 4.

18 Transpose of a Matrix

Given an $m \times n$ matrix A, the **transpose** of A is the $n \times m$ matrix A^T whose entry in row i column j is the entry in row j column i of A. In other words, the rows of A become the columns of A^T, and the columns of A become the rows of A^T.

Example 18.1.

$$A = \begin{bmatrix} 1 & 2 & 8 \\ 0 & -2 & 7 \\ 3 & 1 & 2 \\ 4 & 4 & 1 \end{bmatrix} \quad \Longrightarrow \quad A^T = \begin{bmatrix} 1 & 0 & 3 & 4 \\ 2 & -2 & 1 & 4 \\ 8 & 7 & 2 & 1 \end{bmatrix}$$

$\diamond$

Just as A defines a linear transformation from $\mathbf{R}^n$ to $\mathbf{R}^m$, A^T defines a linear transformation from $\mathbf{R}^m$ to $\mathbf{R}^n$. In this section and the next, we will explore the relationship between these two transformations. First we list some properties of the transpose.

Proposition 18.1. Let A and B be matrices for which the following expressions are defined.

1. $(A^T)^T = A$.

2. $(A + B)^T = A^T + B^T$.

3. $(AB)^T = B^T A^T$.

4. If A is invertible, then A^T is invertible and $(A^T)^{-1} = (A^{-1})^T$.

5. If A is a square matrix, then $\det(A^T) = \det(A)$.

Proof. The first two properties are easy to see. For the third, let $C = AB$, and let $D = B^T A^T$. We will use a_{ij}, b_{ij}, c_{ij} and d_{ij} to denote the entries in row i, column j of A, B, C and D, respectively. Likewise, we denote by a'_{ij}, b'_{ij} and c'_{ij} the entries in row i, column j of A^T, B^T and C^T, respectively. Using this notation, we may write the ij^{th} entry of C^T as

$$c'_{ij} = c_{ji} = \sum_{l=1}^{n} a_{jl} b_{li} = \sum_{l=1}^{n} a'_{lj} b'_{il} = \sum_{l=1}^{n} b'_{il} a'_{lj} = d_{ij}.$$

So $C^T = (AB)^T$ and $D = B^T A^T$ share the same entries, and therefore are equal. To verify the fourth property, let A be an invertible matrix. Taking the transpose of $AA^{-1} = A^{-1}A = I_n$ and using the third property gives

$$(A^{-1})^T A^T = A^T (A^{-1})^T = I_n^T = I_n,$$

so $(A^{-1})^T$ is the inverse of A^T. To prove the equivalence of the determinants we use induction on the size of the matrix. If A is a 2×2 matrix, it is clear that $\det(A^T) = \det(A)$. Now suppose the result holds for all $k \times k$ matrices and suppose A is $(k+1) \times (k+1)$. Expanding the determinant of A^T along column j we have

$$\det(A^T) = \sum_{i=1}^{n} (-1)^{i+j} a'_{ij} \det(A'_{ij}),$$

where, as before, a'_{ij} is the entry in row i column j of A^T and A'_{ij} is the $k \times k$ matrix obtained by removing row i and column j from A^T. But $A'_{ij} = (A_{ji})^T$, so by the induction hypothesis, $\det(A'_{ij}) = \det((A_{ji})^T) = \det(A_{ji})$. Thus, since $a'_{ij} = a_{ji}$,

$$\det(A^T) = \sum_{i=1}^{n} (-1)^{i+j} a_{ji} \det(A_{ji}),$$

which is precisely the expansion of the determinant of A along row j. $\qquad\square$

Dot Products and Transposes

As discussed in Section 7, given a column vector

$$\mathbf{v} = \begin{bmatrix} v_1 \\ \vdots \\ v_n \end{bmatrix}$$

in $\mathbf{R}^n$, its transpose

$$\mathbf{v}^T = \begin{bmatrix} v_1 & \cdots & v_n \end{bmatrix}$$

is a row vector. Multiplying the $1 \times n$ row vector $\mathbf{v}^T$ and the $n \times 1$ column vector $\mathbf{w}$ as matrices gives

$$\mathbf{v}^T \mathbf{w} = \begin{bmatrix} v_1 & \cdots & v_n \end{bmatrix} \begin{bmatrix} w_1 \\ \vdots \\ w_n \end{bmatrix} = v_1 w_1 + \cdots + v_n w_n,$$

precisely the dot product of $\mathbf{v}$ and $\mathbf{w}$. Thus

$$\mathbf{v} \cdot \mathbf{w} = \mathbf{v}^T \mathbf{w}.$$

Given an $m \times n$ matrix A, and a vector $\mathbf{x}$, the product $A\mathbf{x}$ is in $\mathbf{R}^m$, so if $\mathbf{y}$ is a vector in $\mathbf{R}^m$, the dot product $A\mathbf{x} \cdot \mathbf{y}$ is defined. On the other hand, A^T is $n \times m$, so $A^T \mathbf{y}$ is in $\mathbf{R}^n$, and the dot product $\mathbf{x} \cdot A^T \mathbf{y}$ makes sense. Remarkably, these two quantities are always equal.

> **Proposition 18.2.** Let A be an $m \times n$ matrix. Then
>
> $$A\mathbf{x} \cdot \mathbf{y} = \mathbf{x} \cdot A^T \mathbf{y}$$
>
> for all vectors $\mathbf{x}$ in $\mathbf{R}^n$ and $\mathbf{y}$ in $\mathbf{R}^m$.

Proof. Using the expression above for dot products, and the formula for the transpose of a product, we have

$$A\mathbf{x} \cdot \mathbf{y} = (A\mathbf{x})^T \mathbf{y} = \mathbf{x}^T A^T \mathbf{y} = \mathbf{x} \cdot A^T \mathbf{y}.$$

$\square$

Row Space and Left Null Space

Given an $m \times n$ matrix A, we now consider the column space and null space of A^T. Since the columns of A^T are the rows of A, the column space of A^T is the span of the rows of A. We therefore refer to $C(A^T)$ as the **row space** of A. Vectors $\mathbf{x}$ in null space of A^T satisfy $A^T \mathbf{x} = \mathbf{0}$. Taking the transpose of both sides, and viewing the left side as the product of the $n \times m$ matrix A^T with the $m \times 1$ matrix $\mathbf{x}$, we get $\mathbf{x}^T A = \mathbf{0}^T$. This says that for vectors $\mathbf{x}$ in $N(A^T)$, multiplying A on the *left* by the row vector $\mathbf{x}^T$ results in the zero row vector. For this reason, we call $N(A^T)$ the **left null space** of A. So we now have four subspaces associated with the matrix A.

> **The Four Subspaces Associated with an $m \times n$ matrix A**
>
> 1. $N(A)$, the null space of A, a subspace of $\mathbf{R}^n$.
>
> 2. $C(A)$, the column space of A, a subspace of $\mathbf{R}^m$.
>
> 3. $N(A^T)$, the left null space of A, a subspace of $\mathbf{R}^m$.
>
> 4. $C(A^T)$, the row space of A, a subspace of $\mathbf{R}^n$.

Example 18.2. Let

$$A = \begin{bmatrix} 2 & -1 & -3 \\ -4 & 2 & 6 \end{bmatrix}.$$

The four subspaces are

$$N(A) = \text{span}\left(\begin{bmatrix} 1 \\ 2 \\ 0 \end{bmatrix}, \begin{bmatrix} 3 \\ 0 \\ 2 \end{bmatrix} \right)$$

$$C(A) = \text{span}\left(\begin{bmatrix} -1 \\ 2 \end{bmatrix} \right)$$

$$N(A^T) = \text{span}\left(\begin{bmatrix} 2 \\ 1 \end{bmatrix} \right)$$

$$C(A^T) = \text{span}\left(\begin{bmatrix} 2 \\ -1 \\ -3 \end{bmatrix} \right).$$

They are shown in the figure below.

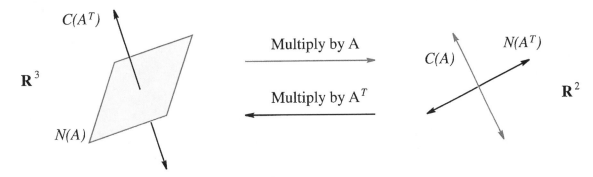

Multiplication by A sends vectors from $\mathbf{R}^3$ into $C(A)$, and vectors from $N(A)$ to the zero vector in $\mathbf{R}^2$, while multiplication by A^T sends vectors from $\mathbf{R}^2$ into $C(A^T)$, and vectors from $N(A^T)$ to the zero vector in $\mathbf{R}^3$. It is interesting to note that the line $C(A)$ is actually perpendicular to the line $N(A^T)$, and the line $C(A^T)$ is perpendicular to the plane $N(A)$. As we shall see in the next section, this is not a coincidence. $\diamond$

We conclude this section by proving that the row space and column space have the same dimension.

Proposition 18.3. Let A be any matrix. Then $\text{rank}(A^T) = \text{rank}(A)$.

Proof. Recall that $\text{rank}(A)$ equals the number of pivots in $\text{rref}(A)$. Since $\text{rank}(A^T)$ is the dimension of the row space of A, let us consider the rows of A and $\text{rref}(A)$. At each stage of the elimination process, the current rows are linear combinations of the rows of A. Hence the rows of $\text{rref}(A)$ are linear combinations of the rows of A, which implies that any vector in the row space of $\text{rref}(A)$ is in the row space of A. By reversing the elimination steps, we can produce A from $\text{rref}(A)$ through a sequence of row operations. Thus, by the same reasoning, any vector in the row space of A is in the row space of $\text{rref}(A)$, and consequently the row space of A equals the row space of $\text{rref}(A)$. It is easy to see that the rows of $\text{rref}(A)$ which contain pivots (the nonzero rows) form a basis for the row space of rref(A), since each such row contains a one in a component where all the other rows contain a zero. This implies that the dimension of the row space of rref(A), and hence the dimension of row space of A, equals the number of pivots in rref(A). $\square$

Exercises

For each matrix A in Exercises 1 through 5 compute A^T, $A^T A$, and $A A^T$.

18.1. $A = \begin{bmatrix} 4 & 2 & 1 \\ 3 & -1 & 5 \end{bmatrix}$

18.2. $A = \begin{bmatrix} 3 & 1 & 2 & 4 \end{bmatrix}$

18.3. $A = \begin{bmatrix} 2 & 1 & 3 \\ 1 & 5 & 4 \\ 3 & 4 & 7 \end{bmatrix}$

18.4. $A = \begin{bmatrix} 5 \\ 1 \\ 0 \\ 2 \end{bmatrix}$

18.5. $A = \begin{bmatrix} 0 & 2 \\ 1 & 3 \\ 6 & 1 \\ 4 & 6 \end{bmatrix}$

For each $m \times n$ matrix A in Exercises 6 through 9, (a) find a basis for $C(A), N(A), C(A^T)$ and $N(A^T)$, (b) graph $N(A)$ and $C(A^T)$ in $\mathbf{R}^n$, $N(A^T)$ and $C(A)$ in $\mathbf{R}^m$. Note that rank(A) = rank(A^T).

18.6. $A = \begin{bmatrix} 1 & 2 \\ 3 & 6 \end{bmatrix}$

18.7. $A = \begin{bmatrix} 1 & 3 & -3 \\ 4 & -2 & 16 \end{bmatrix}$

18.8. $A = \begin{bmatrix} 3 & 4 \\ -6 & 0 \\ 2 & 1 \end{bmatrix}$

18.9. $A = \begin{bmatrix} 3 & 0 & 2 \\ 2 & 4 & -2 \\ 4 & -4 & 6 \end{bmatrix}$

In Exercises 10 through 13, answer true or false. Answer true only if the statement is true for all matrices A, and explain why the statement is true. Otherwise, give a counterexample.

128

18.10. If $S = \{\mathbf{v}_1, \dots, \mathbf{v}_k\}$ is a basis for the row space of $\text{rref}(A)$, then S is a basis for the row space of A.

18.11. If $S = \{\mathbf{v}_1, \dots, \mathbf{v}_k\}$ is a basis for the left null space of $\text{rref}(A)$, then S is a basis for the left null space of A.

18.12. If the left null space of A has dimension d, then the null space of A also has dimension d.

18.13. If $C(A) = \mathbf{R}^m$, then $C(A^T) = \mathbf{R}^n$.

18.14. Let A be an $m \times n$ matrix.

    (a) What is $\text{rank}(A) + \text{nullity}(A)$?

    (b) What is $\text{rank}(A^T) + \text{nullity}(A^T)$?

    (c) Use Proposition 18.3 along with parts (a) and (b) to derive a relationship between $\text{nullity}(A^T)$ and $\text{nullity}(A)$.

19   Orthogonal Complements

Let V be a subspace of $\mathbf{R}^n$. The **orthogonal complement** of V is the set

$$V^\perp = \{\mathbf{x} \in \mathbf{R}^n \mid \mathbf{x} \cdot \mathbf{v} = 0 \text{ for every } \mathbf{v} \text{ in } V\}.$$

That is, $V^\perp$ (read "V perp") consists of all vectors which are orthogonal to every vector in V.

Example 19.1. Let V be the line spanned by

$$\mathbf{v} = \begin{bmatrix} 3 \\ 1 \\ 2 \end{bmatrix}.$$

Any vector

$$\mathbf{x} = \begin{bmatrix} x_1 \\ x_2 \\ x_3 \end{bmatrix}$$

in $V^\perp$ must be orthogonal to $\mathbf{v}$. In addition, $\mathbf{x}$ must be orthogonal to every scalar multiple of $\mathbf{v}$. But if $\mathbf{x}$ is orthogonal to $\mathbf{v}$, orthogonality to scalar multiples of $\mathbf{v}$ comes for free because

$$\mathbf{x} \cdot (c\mathbf{v}) = c(\mathbf{x} \cdot \mathbf{v}) = c0 = 0.$$

Thus $\mathbf{x}$ is in $V^\perp$ if and only if $3x_1 + x_2 + 2x_3 = 0$. So $V^\perp$ is the plane passing through the origin with normal vector $\mathbf{v}$.

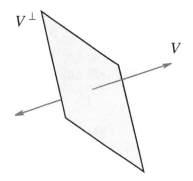

Notice that $V^\perp$ is also a subspace.

◇

Proposition 19.1. Let V be a subspace of $\mathbf{R}^n$. Then $V^\perp$ is also a subspace.

Proof. Since $\mathbf{0} \cdot \mathbf{v} = 0$ for any vector $\mathbf{v}$, and in particular for elements of V, it follows that $\mathbf{0}$ is in $V^\perp$. Next, suppose $\mathbf{x}$ and $\mathbf{y}$ are in $V^\perp$. Then $\mathbf{x} \cdot \mathbf{v} = 0$ and $\mathbf{y} \cdot \mathbf{v} = 0$ for every vector $\mathbf{v}$ in V. It then follows that

$$(\mathbf{x} + \mathbf{y}) \cdot \mathbf{v} = \mathbf{x} \cdot \mathbf{v} + \mathbf{y} \cdot \mathbf{v} = 0 + 0 = 0$$

and

$$(c\mathbf{x}) \cdot \mathbf{v} = c(\mathbf{x} \cdot \mathbf{v}) = c0 = 0$$

for every vector $\mathbf{v}$ in V, so both $\mathbf{x} + \mathbf{y}$ and $c\mathbf{x}$ are in $V^\perp$. □

We now state the relationship between the four subspaces associated with a matrix A.

Proposition 19.2. Let A be any matrix.

1. $N(A) = C(A^T)^\perp$.

2. $N(A^T) = C(A)^\perp$.

So the null space consists of all vectors which are orthogonal to everything in the row space, and likewise the left null space consists of all vectors orthogonal to everything in the column space.

Proof. Write A as

$$A = \begin{bmatrix} \text{---} & \mathbf{w}_1^T & \text{---} \\ \text{---} & \mathbf{w}_2^T & \text{---} \\ & \vdots & \\ \text{---} & \mathbf{w}_m^T & \text{---} \end{bmatrix}$$

in terms of its rows. If $\mathbf{x}$ is in the null space of A, then

$$A\mathbf{x} = \begin{bmatrix} \mathbf{w}_1 \cdot \mathbf{x} \\ \mathbf{w}_2 \cdot \mathbf{x} \\ \vdots \\ \mathbf{w}_m \cdot \mathbf{x} \end{bmatrix} = \begin{bmatrix} 0 \\ 0 \\ \vdots \\ 0 \end{bmatrix}$$

so $\mathbf{x}$ is orthogonal to each row of A. By linearity of the dot product,

$$(c_1\mathbf{w}_1 + c_2\mathbf{w}_2 + \cdots + c_m\mathbf{w}_m) \cdot \mathbf{x} = c_1\mathbf{w}_1 \cdot \mathbf{x} + c_2\mathbf{w}_2 \cdot \mathbf{x} + \cdots + c_m\mathbf{w}_m \cdot \mathbf{x}$$
$$= 0$$

so $\mathbf{x}$ is orthogonal to any vector in the row space of A. Hence $\mathbf{x}$ is in $C(A^T)^\perp$. Conversely, suppose $\mathbf{x}$ is some vector in $C(A^T)^\perp$. Then $\mathbf{x}$ is orthogonal to every vector in the row space of A. In particular, $\mathbf{x}$ is orthogonal to each row of A, so $A\mathbf{x} = \mathbf{0}$, and therefore $\mathbf{x}$ is in the null space of A. Thus $C(A^T)^\perp = N(A)$. The second statement follows by replacing A with A^T (and hence A^T with $(A^T)^T = A$) in the first statement. $\qquad\square$

An important consequence of this is that a subspace and its orthogonal complement have complementary dimensions.

Proposition 19.3. Let V be a subspace of $\mathbf{R}^n$. Then $\dim(V) + \dim(V^\perp) = n$.

Proof. If $V = \{\mathbf{0}\}$, then $V^\perp = \mathbf{R}^n$, and the statement holds. Otherwise, let $\{\mathbf{v}_1, \ldots, \mathbf{v}_k\}$ be a basis for V, and let A be the matrix whose columns consist of these basis vectors. Then $V = C(A)$, so $V^\perp = C(A)^\perp = N(A^T)$, by Proposition 19.2. The matrix A^T is $k \times n$, so by the Rank-Nullity Theorem, $\dim(C(A^T)) + \dim(N(A^T)) = n$. By Proposition 18.3, $\dim(C(A^T)) = \dim(C(A))$, so $\dim(C(A)) + \dim(N(A^T)) = n$. $\qquad\square$

Proposition 19.4. Let V be a subspace of $\mathbf{R}^n$.

1. $V \cap V^\perp = \{\mathbf{0}\}$.

2. $V + V^\perp = \mathbf{R}^n$. That is, every vector $\mathbf{x}$ in $\mathbf{R}^n$ can be written as a sum $\mathbf{v} + \mathbf{w}$ of a vector $\mathbf{v}$ in V and a vector $\mathbf{w}$ in $V^\perp$. Furthermore, this expression is unique.

Proof. To prove the first statement, suppose $\mathbf{v}$ is in both V and $V^\perp$. Then $\mathbf{v}$ is orthogonal to itself, so $\|\mathbf{v}\|^2 = \mathbf{v} \cdot \mathbf{v} = 0$, which implies $\mathbf{v} = \mathbf{0}$.

If $V = \{\mathbf{0}\}$, then $V^\perp = \mathbf{R}^n$, so $\mathbf{x} = \mathbf{0} + \mathbf{x}$ is a sum of a vector in V and a vector in $V^\perp$. If $V = \mathbf{R}^n$, then $V^\perp = \{\mathbf{0}\}$, so $\mathbf{x} = \mathbf{x} + \mathbf{0}$ is a sum of a vector in V and a vector in $V^\perp$. Otherwise, V has dimension $0 < k < n$ and $V^\perp$ has dimension $0 < n - k < n$, so let $\{\mathbf{v}_1, \ldots, \mathbf{v}_k\}$ be a basis for V, let $\{\mathbf{w}_1, \ldots, \mathbf{w}_{n-k}\}$ be a basis for $V^\perp$, and suppose that

$$c_1\mathbf{v}_1 + \cdots + c_k\mathbf{v}_k + d_1\mathbf{w}_1 + \cdots + d_{n-k}\mathbf{w}_{n-k} = \mathbf{0}$$

131

for some scalars. Moving the vectors in $V^\perp$ to the right side, we have

$$\mathbf{v} = c_1\mathbf{v}_1 + \cdots + c_k\mathbf{v}_k = -(d_1\mathbf{w}_1 + \cdots + d_{n-k}\mathbf{w}_{n-k}).$$

Since $\mathbf{v}$ is in both V and $V^\perp$, it equals $\mathbf{0}$ by the first statement, so

$$c_1\mathbf{v}_1 + \cdots + c_k\mathbf{v}_k = \mathbf{0}$$

and

$$d_1\mathbf{w}_1 + \cdots + d_{n-k}\mathbf{w}_{n-k} = \mathbf{0}.$$

By independence, c_1 through c_k and d_1 through d_{n-k} must all be zero. So the set

$$\{\mathbf{v}_1, \ldots, \mathbf{v}_k, \mathbf{w}_1, \ldots, \mathbf{w}_{n-k}\}$$

is linearly independent, and since it consists of n vectors, by Proposition 12.3 it is a basis for $\mathbf{R}^n$. Any vector $\mathbf{x}$ in $\mathbf{R}^n$ may then be written

$$\mathbf{x} = \underbrace{c_1\mathbf{v}_1 + \cdots + c_k\mathbf{v}_k}_{\text{in } V} + \underbrace{d_1\mathbf{w}_1 + \cdots + d_{n-k}\mathbf{w}_{n-k}}_{\text{in } V^\perp}.$$

To see that such expressions are unique, suppose $\mathbf{x} = \mathbf{v}_1 + \mathbf{w}_1 = \mathbf{v}_2 + \mathbf{w}_2$, where $\mathbf{v}_1$ and $\mathbf{v}_2$ are in V and $\mathbf{w}_1$ and $\mathbf{w}_2$ are in $V^\perp$. Then $\mathbf{v}_1 - \mathbf{v}_2 = \mathbf{w}_2 - \mathbf{w}_1$ is a vector in both V and $V^\perp$, and is therefore the zero vector. So $\mathbf{v}_1 = \mathbf{v}_2$ and $\mathbf{w}_1 = \mathbf{w}_2$. □

Proposition 19.5. $(V^\perp)^\perp = V$.

Proof. First let $\mathbf{x}$ be a vector in $(V^\perp)^\perp$. By Proposition 19.4, $\mathbf{x} = \mathbf{v} + \mathbf{w}$ for some $\mathbf{v}$ in V and $\mathbf{w}$ in $V^\perp$. Since $\mathbf{x}$ is orthogonal to $\mathbf{w}$ and $\mathbf{w}$ is orthogonal to $\mathbf{v}$, taking the dot product of both sides with $\mathbf{w}$ gives

$$0 = \mathbf{x} \cdot \mathbf{w} = \mathbf{v} \cdot \mathbf{w} + \mathbf{w} \cdot \mathbf{w} = \|\mathbf{w}\|^2.$$

Thus $\mathbf{w} = \mathbf{0}$, and $\mathbf{x} = \mathbf{v}$ is in V. Conversely, suppose $\mathbf{v}$ is a vector in V. Applying Proposition 19.4 to the subspace $V^\perp$, we may write $\mathbf{v} = \mathbf{w} + \mathbf{x}$ for some $\mathbf{w}$ in $V^\perp$ and some $\mathbf{x}$ in $(V^\perp)^\perp$. Again taking the dot product with $\mathbf{w}$ gives

$$0 = \mathbf{v} \cdot \mathbf{w} = \mathbf{w} \cdot \mathbf{w} + \mathbf{x} \cdot \mathbf{w} = \|\mathbf{w}\|^2.$$

So $\mathbf{w} = \mathbf{0}$ and $\mathbf{v} = \mathbf{x}$ is in $(V^\perp)^\perp$. □

The importance of this result is that if a subspace W is the orthogonal complement of a subspace V, then $W = V^\perp$, so by Proposition 19.5, $W^\perp = (V^\perp)^\perp = V$, and therefore V is the orthogonal complement of W. Thus there is a symmetry in the definition. For instance, according to Proposition 19.2 we have

1. $C(A^T) = N(A)^\perp$.

2. $C(A) = N(A^T)^\perp$.

Example 19.2. Find the set V of all vectors in $\mathbf{R}^4$ which are orthogonal to every solution of

$$
\begin{array}{rcrcrcrcl}
2x_1 & + & x_2 & - & 3x_3 & + & x_4 & = & 0 \\
-2x_1 & + & 3x_2 & - & x_3 & + & 2x_4 & = & 0.
\end{array}
$$

Solution. The set of solutions is the null space of the matrix

$$
A = \begin{bmatrix} 2 & 1 & -3 & 1 \\ -2 & 3 & -1 & 2 \end{bmatrix}
$$

so the set of vectors we are seeking is $V = N(A)^\perp = C(A^T)$, the row space of A. The vectors

$$
\begin{bmatrix} 2 \\ 1 \\ -3 \\ 1 \end{bmatrix} \quad \text{and} \quad \begin{bmatrix} -2 \\ 3 \\ -1 \\ 2 \end{bmatrix}
$$

form a basis for V. $\diamond$

We conclude this section by illustrating further the relationship between the column space and row space of a matrix.

Proposition 19.6. Given any vector $\mathbf{b}$ in the column space of A, there exists a unique vector $\mathbf{x}_0$ in the row space of A such that $A\mathbf{x}_0 = \mathbf{b}$. Furthermore, among all solutions of $A\mathbf{x} = \mathbf{b}$, $\mathbf{x}_0$ has the smallest magnitude.

Proof. For any $\mathbf{b}$ in $C(A)$, there is at least one solution $\mathbf{x}$ of $A\mathbf{x} = \mathbf{b}$. By Proposition 19.4, $\mathbf{x} = \mathbf{x}_0 + \mathbf{y}$ for some $\mathbf{x}_0$ in $C(A^T)$ and some $\mathbf{y}$ in $C(A^T)^\perp = N(A)$. Thus

$$
A\mathbf{x}_0 = A(\mathbf{x} - \mathbf{y}) = A\mathbf{x} - A\mathbf{y} = \mathbf{b} - \mathbf{0} = \mathbf{b},
$$

so $\mathbf{x}_0$ is an element of the row space which satisfies $A\mathbf{x}_0 = \mathbf{b}$. To prove that this is the only such vector in the row space, suppose $\mathbf{x}_1$ is in the row space and satisfies $A\mathbf{x}_1 = \mathbf{b}$. Then since $C(A^T)$ is a subspace, $\mathbf{x}_1 - \mathbf{x}_0$ is in $C(A^T)$, and since

$$
A(\mathbf{x}_1 - \mathbf{x}_0) = A\mathbf{x}_1 - A\mathbf{x}_0 = \mathbf{b} - \mathbf{b} = \mathbf{0},
$$

$\mathbf{x}_1 - \mathbf{x}_0$ is in $N(A) = C(A^T)^\perp$. Thus by the first part of Proposition 19.4, $\mathbf{x}_1 - \mathbf{x}_0 = \mathbf{0}$, so $\mathbf{x}_1 = \mathbf{x}_0$, which proves uniqueness.

For any solution $\mathbf{x}$ of $A\mathbf{x} = \mathbf{b}$, we see as above that $\mathbf{x} - \mathbf{x}_0$ is in $N(A) = C(A^T)^\perp$ and is therefore orthogonal to $\mathbf{x}_0$. Hence, by the Pythagorean Theorem,

$$
\|\mathbf{x}\|^2 = \|(\mathbf{x} - \mathbf{x}_0) + \mathbf{x}_0\|^2 = \|\mathbf{x} - \mathbf{x}_0\|^2 + \|\mathbf{x}_0\|^2 \geq \|\mathbf{x}_0\|^2,
$$

so $\|\mathbf{x}_0\| \leq \|\mathbf{x}\|$. $\square$

Example 19.3. Let

$$A = \begin{bmatrix} 3 & -2 \\ 6 & -4 \end{bmatrix} \qquad \text{and} \qquad \mathbf{b} = \begin{bmatrix} 9 \\ 18 \end{bmatrix}.$$

Then

$$N(A) = \text{span}\left(\begin{bmatrix} 2 \\ 3 \end{bmatrix}\right), \qquad C(A^T) = \text{span}\left(\begin{bmatrix} 3 \\ -2 \end{bmatrix}\right),$$

and the set of solutions of $A\mathbf{x} = \mathbf{b}$ is

$$\left\{ \begin{bmatrix} 3 \\ 0 \end{bmatrix} + t \begin{bmatrix} 2 \\ 3 \end{bmatrix} \;\middle|\; t \in \mathbf{R} \right\}.$$

To find the solution $\mathbf{x}_0$ in $C(A^T)$, we use the fact that, for any solution $\mathbf{x}$, the difference $\mathbf{x} - \mathbf{x}_0$ in orthogonal to the line $L = C(A^T)$. Thus $\mathbf{x}_0$ must be the orthogonal projection of $\mathbf{x}$ onto $C(A^T)$. Choosing any solution, say

$$\mathbf{x} = \begin{bmatrix} 3 \\ 0 \end{bmatrix},$$

and letting

$$\mathbf{u} = \frac{1}{13} \begin{bmatrix} 3 \\ -2 \end{bmatrix}$$

denote one of the unit vectors which spans $C(A^T)$, we then find that

$$\mathbf{x}_0 = \mathbf{Proj}_L(\mathbf{x}) = (\mathbf{x} \cdot \mathbf{u})\mathbf{u} = \begin{bmatrix} 27/13 \\ -18/13 \end{bmatrix}.$$

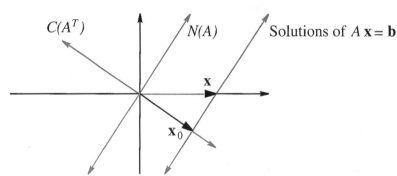

This solution $\mathbf{x}_0$ has the smallest magnitude among all solutions of $A\mathbf{x} = \mathbf{b}$. It is closest to the origin, and the only solution orthogonal to the null space of A. $\diamond$

The solution $\mathbf{x}_0$ will generally be the orthogonal projection of any solution $\mathbf{x}$ onto the row space. In the next section we consider orthogonal projections onto arbitrary subspaces.

Exercises

In Exercises 1 through 16, for the subspace V given, find a basis for $V^\perp$.

19.1. $V = \mathrm{span}\left(\begin{bmatrix} 2 \\ 5 \end{bmatrix} \right)$

19.2. $V = \mathrm{span}\left(\begin{bmatrix} 2 \\ 5 \end{bmatrix}, \begin{bmatrix} 1 \\ 4 \end{bmatrix} \right)$

19.3. $V = \mathrm{span}\left(\begin{bmatrix} 3 \\ -2 \\ 1 \end{bmatrix} \right)$

19.4. $V = \mathrm{span}\left(\begin{bmatrix} 2 \\ 0 \\ 4 \end{bmatrix}, \begin{bmatrix} 3 \\ -2 \\ 1 \end{bmatrix} \right)$

19.5. $V = \mathrm{span}\left(\begin{bmatrix} 2 \\ 0 \\ 4 \end{bmatrix}, \begin{bmatrix} 3 \\ -2 \\ 1 \end{bmatrix}, \begin{bmatrix} -5 \\ 4 \\ 0 \end{bmatrix} \right)$

19.6. $V = \mathrm{span}\left(\begin{bmatrix} 2 \\ 0 \\ 4 \end{bmatrix}, \begin{bmatrix} 3 \\ -2 \\ 1 \end{bmatrix}, \begin{bmatrix} 5 \\ 2 \\ 1 \end{bmatrix} \right)$

19.7. $V = \mathrm{span}\left(\begin{bmatrix} 5 \\ 2 \\ -2 \\ 1 \end{bmatrix} \right)$

19.8. $V = \mathrm{span}\left(\begin{bmatrix} 5 \\ 2 \\ -2 \\ 1 \end{bmatrix}, \begin{bmatrix} 2 \\ 3 \\ 0 \\ 5 \end{bmatrix} \right)$

19.9. V is the line in $\mathbf{R}^2$ which satisfies the equation $-x + 7y = 0$.

19.10. V is the plane in $\mathbf{R}^3$ which satisfies the equation $4x - 2y + z = 0$.

19.11. V is the hyperplane in $\mathbf{R}^4$ given by $3x + 2y - z + 7w = 0$.

19.12. V is the set of solutions to the system

$$
\begin{aligned}
3x_1 &+ 2x_2 = 0 \\
-x_1 &+ 5x_2 = 0
\end{aligned}
$$

19.13. V is the set of solutions to the system

$$
\begin{aligned}
3x_1 &- x_2 + 7x_3 = 0 \\
2x_1 &+ 4x_2 + x_3 = 0
\end{aligned}
$$

19.14. V is the set of solutions to the system

$$\begin{array}{rcrcrcr} x_1 & - & 3x_2 & + & 4x_3 & = & 0 \\ x_1 & + & 6x_2 & - & 5x_3 & = & 0 \\ 4x_1 & + & 6x_2 & - & 2x_3 & = & 0 \end{array}$$

19.15. V is the set of solutions to the system

$$\begin{array}{rcrcrcrcr} 3x_1 & - & 7x_2 & + & 4x_3 & + & x_4 & = & 0 \\ 2x_1 & & & + & 5x_3 & - & 4x_4 & = & 0 \end{array}$$

19.16. V is the set of solutions to the system

$$\begin{array}{rcrcrcrcr} -x_1 & + & 3x_2 & + & 4x_3 & - & 5x_4 & = & 0 \\ 4x_1 & - & 6x_2 & + & 11x_3 & - & 4x_4 & = & 0 \\ 2x_1 & - & 4x_2 & + & x_3 & + & 2x_4 & = & 0 \end{array}$$

In Exercises 17 through 22, find a basis for each of the four subspaces. Notice that $N(A) = C(A^T)^{\perp}$ and $N(A^T) = C(A)^{\perp}$.

19.17. $A = \begin{bmatrix} 2 & 3 & -1 \\ 4 & 0 & 2 \end{bmatrix}$

19.18. $A = \begin{bmatrix} 4 & 2 \\ -3 & 1 \\ 2 & 5 \end{bmatrix}$

19.19. $A = \begin{bmatrix} 2 & -1 & 3 \\ 4 & 2 & 0 \\ 8 & -2 & 9 \end{bmatrix}$

19.20. $A = \begin{bmatrix} 1 & 0 & 4 & -5 \\ 2 & 2 & -5 & 9 \\ 3 & 2 & -1 & 4 \\ 8 & 6 & -7 & 17 \end{bmatrix}$

19.21. $A = \begin{bmatrix} 1 & 1 & 0 & 1 \\ 0 & 1 & 1 & 1 \\ 1 & 1 & 1 & 0 \\ 2 & 1 & 0 & 0 \end{bmatrix}$

19.22. $A = \begin{bmatrix} 2 & -1 & 0 & 5 \\ 3 & 8 & 1 & -4 \\ 3 & -11 & -1 & 19 \end{bmatrix}$

19.23. Let V, W be subspaces of $\mathbf{R}^n$. Show that if $V \subseteq W$, then $W^{\perp} \subseteq V^{\perp}$.

19.24. For any subspaces X and Y define $X+Y \equiv \{\mathbf{u} = \mathbf{x}+\mathbf{y} : \mathbf{x} \in X, \mathbf{y} \in Y\}$. Let V, W be subspaces of $\mathbf{R}^n$. Use the following steps to show that $(V \cap W)^\perp = V^\perp + W^\perp$.

(a) Show that $V \cap W \subseteq (V^\perp + W^\perp)^\perp$.

(b) Show that $(V^\perp + W^\perp)^\perp \subseteq V \cap W$. *Hint:* Use the result from Exercise 23 to show that $(V^\perp + W^\perp)^\perp \subseteq V$ and $(V^\perp + W^\perp)^\perp \subseteq W$.

(c) Use the fact that for any subspace U, $(U^\perp)^\perp = U$ to conclude that $(V \cap W)^\perp = V^\perp + W^\perp$.

19.25. Let V, W be subspaces of $\mathbf{R}^n$. Use the following steps to show that $(V \cup W)^\perp = V^\perp \cap W^\perp$.

(a) $(V \cup W)^\perp \subseteq V^\perp \cap W^\perp$.

(b) $V^\perp \cap W^\perp \subseteq (V \cup W)^\perp$.

19.26. Suppose

$$\left\{ \begin{bmatrix} 3 \\ -1 \\ 2 \end{bmatrix}, \begin{bmatrix} 1 \\ 1 \\ -1 \end{bmatrix} \right\}$$

is a basis for the left null space of A. For which vectors $\mathbf{b}$ does the system $A\mathbf{x} = \mathbf{b}$ have a solution?

(a) $\mathbf{b} = \begin{bmatrix} 2 \\ -10 \\ -8 \end{bmatrix}$
(b) $\mathbf{b} = \begin{bmatrix} 3 \\ -1 \\ 2 \end{bmatrix}$
(c) $\mathbf{b} = \begin{bmatrix} 4 \\ 0 \\ 1 \end{bmatrix}$
(d) $\mathbf{b} = \begin{bmatrix} -1 \\ 5 \\ 4 \end{bmatrix}$

19.27. Suppose

$$\left\{ \begin{bmatrix} 1 \\ 2 \\ 3 \end{bmatrix} \right\}$$

is a basis for the left null space of A. For which vectors $\mathbf{b}$ does the system $A\mathbf{x} = \mathbf{b}$ have a solution?

(a) $\mathbf{b} = \begin{bmatrix} 1 \\ 1 \\ -1 \end{bmatrix}$
(b) $\mathbf{b} = \begin{bmatrix} 1 \\ 2 \\ 3 \end{bmatrix}$
(c) $\mathbf{b} = \begin{bmatrix} -1 \\ 5 \\ -3 \end{bmatrix}$
(d) $\mathbf{b} = \begin{bmatrix} 3 \\ -2 \\ 1 \end{bmatrix}$

19.28. Let A be an $m \times n$ matrix and suppose that the system $A\mathbf{x} = \mathbf{b}$ has a unique solution for some $\mathbf{b}$ in $\mathbf{R}^m$. What does this imply about the row space of A?

In Exercises 26 through 29 find the solution of $A\mathbf{x} = \mathbf{b}$ which lies in the row space of A.

19.26. $A = \begin{bmatrix} 1 & -2 & 3 \\ 2 & -4 & 6 \end{bmatrix}$, $\mathbf{b} = \begin{bmatrix} 3 \\ 6 \end{bmatrix}$

$19.27.$ $A = \begin{bmatrix} 3 & -6 \\ -2 & 4 \end{bmatrix}$, $\quad \mathbf{b} = \begin{bmatrix} -3 \\ 2 \end{bmatrix}$

$19.28.$ $A = \begin{bmatrix} 5 & 2 \\ -5 & -2 \end{bmatrix}$, $\quad \mathbf{b} = \begin{bmatrix} -3 \\ 3 \end{bmatrix}$

$19.29.$ $A = \begin{bmatrix} 1 & 2 & -1 & 3 \\ -1 & -2 & 1 & -3 \\ 2 & 4 & -2 & 6 \end{bmatrix}$, $\quad \mathbf{b} = \begin{bmatrix} 2 \\ -2 \\ 4 \end{bmatrix}$

20 Orthogonal Projections

Let V be a subspace of $\mathbf{R}^n$. By Proposition 19.4 any vector $\mathbf{x}$ in $\mathbf{R}^n$ may be written uniquely as a sum

$$\mathbf{x} = \mathbf{v} + \mathbf{w}$$

where $\mathbf{v}$ is in V and $\mathbf{w}$ is in $V^\perp$. We can therefore define a function $\mathbf{Proj}_V : \mathbf{R}^n \to \mathbf{R}^n$, called the **projection** onto the subspace V, by $\mathbf{Proj}_V(\mathbf{x}) = \mathbf{v}$.

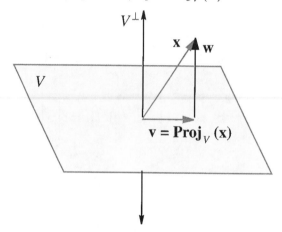

We claim that $\mathbf{Proj}_V$ is a linear transformation. We will show this in the case that V is a nontrivial proper subspace of $\mathbf{R}^n$ – the cases $V = \{\mathbf{0}\}$ and $V = \mathbf{R}^n$ are left for the reader to consider. Let $\{\mathbf{v}_1, \ldots, \mathbf{v}_k\}$ be a basis for V and let A be the $n \times k$ matrix whose columns are these vectors. Then $V = C(A)$, so that $\mathbf{Proj}_V(\mathbf{x}) = A\mathbf{y}$ for some vector $\mathbf{y}$ in $\mathbf{R}^k$. By definition, $\mathbf{x} - \mathbf{Proj}_V(\mathbf{x})$ is a vector $\mathbf{w}$ in $V^\perp = C(A)^\perp$. But by Proposition 19.2, $C(A)^\perp = N(A^T)$, so $\mathbf{x} - A\mathbf{y}$ is in the null space of A^T. Thus

$$A^T(\mathbf{x} - A\mathbf{y}) = \mathbf{0} \qquad \Longrightarrow \qquad A^T A\mathbf{y} = A^T\mathbf{x}.$$

It turns out that the $k \times k$ matrix $A^T A$ is invertible. To see this, suppose that $\mathbf{z}$ is in the null space of $A^T A$. Then $A^T A\mathbf{z} = \mathbf{0}$. Multiplying on the left by $\mathbf{z}^T$ gives $\mathbf{z}^T A^T A\mathbf{z} = 0$. Since $\mathbf{z}^T A^T = (A\mathbf{z})^T$, this implies that $A\mathbf{z} \cdot A\mathbf{z} = 0$, and therefore $\|A\mathbf{z}\|^2 = 0$. Since only the zero vector has length zero, this implies $A\mathbf{z} = \mathbf{0}$. But since the columns of A are linearly independent, $N(A) = \{\mathbf{0}\}$, so $\mathbf{z} = \mathbf{0}$. Hence the null space of $A^T A$ is trivial, and since $A^T A$ is a square matrix, it must be invertible. We may therefore solve the above equation for $\mathbf{y}$ to get $\mathbf{y} = (A^T A)^{-1} A^T \mathbf{x}$ and therefore

$$\mathbf{Proj}_V(\mathbf{x}) = A(A^TA)^{-1}A^T\mathbf{x}.$$

Example 20.1. Find the matrix for the projection onto the subspace

$$V = \operatorname{span}\left(\begin{bmatrix}1\\0\\0\\1\end{bmatrix}, \begin{bmatrix}0\\1\\0\\1\end{bmatrix}\right)$$

of $\mathbf{R}^4$.
Solution.

$$A = \begin{bmatrix}1 & 0\\0 & 1\\0 & 0\\1 & 1\end{bmatrix} \quad \Longrightarrow \quad A^TA = \begin{bmatrix}2 & 1\\1 & 2\end{bmatrix} \quad \Longrightarrow \quad (A^TA)^{-1} = \frac{1}{3}\begin{bmatrix}2 & -1\\-1 & 2\end{bmatrix}$$

so

$$A(A^TA)^{-1}A^T = \frac{1}{3}\begin{bmatrix}1 & 0\\0 & 1\\0 & 0\\1 & 1\end{bmatrix}\begin{bmatrix}2 & -1\\-1 & 2\end{bmatrix}\begin{bmatrix}1 & 0 & 0 & 1\\0 & 1 & 0 & 1\end{bmatrix} = \frac{1}{3}\begin{bmatrix}2 & -1 & 0 & 1\\-1 & 2 & 0 & 1\\0 & 0 & 0 & 0\\1 & 1 & 0 & 2\end{bmatrix}.$$

$\diamond$

The following result states that the projections onto V and $V^\perp$ are complementary, in that they sum to the identity function.

Proposition 20.1. Let V be a subspace of $\mathbf{R}^n$. Then $\mathbf{Proj}_V + \mathbf{Proj}_{V^\perp} = \mathbf{I}_{\mathbf{R}^n}$.

Proof. Applying Proposition 19.4 to $V^\perp$, we can express any vector $\mathbf{x}$ in $\mathbf{R}^n$ uniquely as $\mathbf{w}+\mathbf{v}$ where $\mathbf{w}$ is in $V^\perp$ and $\mathbf{v}$ is in $(V^\perp)^\perp = V$. By definition $\mathbf{w} = \mathbf{Proj}_{V^\perp}(\mathbf{x})$ and $\mathbf{v} = \mathbf{Proj}_V(\mathbf{x})$, so $\mathbf{x} = \mathbf{Proj}_V(\mathbf{x}) + \mathbf{Proj}_{V^\perp}(\mathbf{x})$. $\square$

Example 20.2. Find the matrix for the projection onto the plane $x_1 + x_2 + x_3 = 0$ in $\mathbf{R}^3$.
Solution. Call the plane V. Then V is the null space of the matrix $A = \begin{bmatrix}1 & 1 & 1\end{bmatrix}$, and thus $V^\perp = N(A)^\perp = C(A^T)$ is the row space of A. That is,

$$V^\perp = \operatorname{span}\left(\begin{bmatrix}1\\1\\1\end{bmatrix}\right).$$

139

Call B the matrix consisting of this single column vector. Then

$$B(B^T B)^{-1} B^T = \begin{bmatrix} 1 \\ 1 \\ 1 \end{bmatrix} \begin{bmatrix} 3 \end{bmatrix}^{-1} \begin{bmatrix} 1 & 1 & 1 \end{bmatrix} = \frac{1}{3} \begin{bmatrix} 1 & 1 & 1 \\ 1 & 1 & 1 \\ 1 & 1 & 1 \end{bmatrix}$$

is the matrix for $\mathbf{Proj}_{V^{\perp}}$. Since $\mathbf{Proj}_V = I_{\mathbf{R}^3} - \mathbf{Proj}_{V^{\perp}}$, the matrix for $\mathbf{Proj}_V$ is

$$\begin{bmatrix} 1 & 0 & 0 \\ 0 & 1 & 0 \\ 0 & 0 & 1 \end{bmatrix} - \frac{1}{3} \begin{bmatrix} 1 & 1 & 1 \\ 1 & 1 & 1 \\ 1 & 1 & 1 \end{bmatrix} = \frac{1}{3} \begin{bmatrix} 2 & -1 & -1 \\ -1 & 2 & -1 \\ -1 & -1 & 2 \end{bmatrix}.$$

$\diamond$

Geometrically, the orthogonal projection of a vector $\mathbf{x}$ onto a subspace V is the closest point in V to the point $\mathbf{x}$.

> **Proposition 20.2.** Let V be a subspace of $\mathbf{R}^n$, and let $\mathbf{x}$ be any vector in $\mathbf{R}^n$. Then
>
> $$\|\mathbf{x} - \mathbf{Proj}_V(\mathbf{x})\| \leq \|\mathbf{x} - \mathbf{v}\|$$
>
> for every $\mathbf{v}$ in V, and equality holds if and only if $\mathbf{v} = \mathbf{Proj}_V(\mathbf{x})$.

Proof. Let $\mathbf{v}$ be any vector in V and write

$$\mathbf{x} - \mathbf{v} = \underbrace{\mathbf{x} - \mathbf{Proj}_V(\mathbf{x})}_{\mathbf{a}} + \underbrace{\mathbf{Proj}_V(\mathbf{x}) - \mathbf{v}}_{\mathbf{b}}.$$

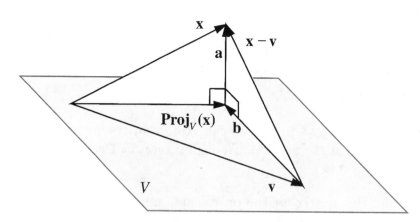

By definition $\mathbf{a} = \mathbf{x} - \mathbf{Proj}_V(\mathbf{x})$ is in $V^{\perp}$, and since V is a subspace, $\mathbf{b} = \mathbf{Proj}_V(\mathbf{x}) - \mathbf{v}$ is in V. So $\mathbf{a}$ and $\mathbf{b}$ are orthogonal, and the Pythagorean Theorem implies

$$\begin{aligned} \|\mathbf{x} - \mathbf{v}\|^2 &= \|\mathbf{a} + \mathbf{b}\|^2 \\ &= \|\mathbf{a}\|^2 + \|\mathbf{b}\|^2 \\ &= \|\mathbf{x} - \mathbf{Proj}_V(\mathbf{x})\|^2 + \|\mathbf{Proj}_V(\mathbf{x}) - \mathbf{v}\|^2. \end{aligned}$$

Hence

$$\|\mathbf{x} - \mathbf{Proj}_V(\mathbf{x})\| \le \|\mathbf{x} - \mathbf{v}\|$$

and we have equality only if $\mathbf{v} = \mathbf{Proj}_V(\mathbf{x})$. $\qquad\square$

Example 20.3. Find the distance between the point $(3, 2, 1)$ and the plane $x_1 + x_2 + x_3 = 0$.
Solution. Using the matrix for the orthogonal projection onto the plane found in Example 20.2, we have

$$\frac{1}{3}\begin{bmatrix} 2 & -1 & -1 \\ -1 & 2 & -1 \\ -1 & -1 & 2 \end{bmatrix}\begin{bmatrix} 3 \\ 2 \\ 1 \end{bmatrix} = \begin{bmatrix} 1 \\ 0 \\ -1 \end{bmatrix}.$$

Thus $(1, 0, -1)$ is the closest point in the plane to $(3, 2, 1)$, and the distance between these points is $2\sqrt{3}$. $\qquad\diamond$

Least Squares Approximation

Next suppose we seek a solution to a linear system $A\mathbf{x} = \mathbf{b}$, but the right hand side $\mathbf{b}$ is not in the column space of A, so that the system has no solutions. To obtain an approximate solution, we attempt to find an $\mathbf{x}$ which minimizes the length of the difference vector $A\mathbf{x} - \mathbf{b}$. For such an $\mathbf{x}$, by Proposition 20.2, $A\mathbf{x}$ must be the projection of $\mathbf{b}$ onto $V = C(A)$.

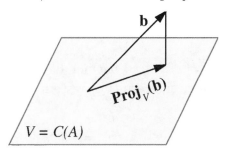

Thus we seek a solution $\mathbf{x}^*$ of $A\mathbf{x}^* = \mathbf{Proj}_V(\mathbf{b})$. The right hand side is in $C(A)$, so this system certainly has a solution. Now $\mathbf{x}^*$ is a solution if and only if

$$A\mathbf{x}^* - \mathbf{b} = \mathbf{Proj}_V(\mathbf{b}) - \mathbf{b}$$

is in $V^\perp = C(A)^\perp = N(A^T)$. Thus $\mathbf{x}^*$ is a solution if and only if $A^T(A\mathbf{x}^* - \mathbf{b}) = \mathbf{0}$. This equation simplifies conveniently to

$$A^T A\mathbf{x}^* = A^T\mathbf{b}$$

which is just A^T multiplied by the original system $A\mathbf{x} = \mathbf{b}$. A solution $\mathbf{x}^*$ of this system is called a **least-squares** solution of the system $A\mathbf{x} = \mathbf{b}$. Although this system always has a solution, we emphasize that solutions of this system are *not* solutions of the original system $A\mathbf{x} = \mathbf{b}$, which we assumed had no solutions. A least-squares solution minimizes the error $\|A\mathbf{x} - \mathbf{b}\|$. The term least-squares comes from the fact that the error (squared) is the sum of the squares of the error in each equation of the system.

Example 20.4. The system

$$\begin{array}{rcrcl} 2x_1 & - & x_2 & = & 2 \\ x_1 & + & 2x_2 & = & 1 \\ x_1 & + & x_2 & = & 4 \end{array}$$

has no solutions. (Check!) To find the least squares solution, notice that

$$A^T A = \begin{bmatrix} 6 & 1 \\ 1 & 6 \end{bmatrix} \quad \text{and} \quad A^T \mathbf{b} = \begin{bmatrix} 9 \\ 4 \end{bmatrix}$$

so the system $A^T A \mathbf{x}^* = A^T \mathbf{b}$ takes the form

$$\begin{array}{rcrcl} 6x_1^* & + & x_2^* & = & 9 \\ x_1^* & + & 6x_2^* & = & 4. \end{array}$$

The solution of this system is

$$\mathbf{x}^* = \begin{bmatrix} 10/7 \\ 3/7 \end{bmatrix},$$

and the error is

$$\|A\mathbf{x}^* - \mathbf{b}\| = \left\| \begin{bmatrix} 3/7 \\ 9/7 \\ -15/7 \end{bmatrix} \right\| = 3\sqrt{35}/7.$$

$\diamond$

Example 20.5. To find the line $f(x) = ax + b$ which best fits the data $f(-1) = 0$, $f(0) = 1$, $f(1) = 2$, and $f(2) = 1$, we consider the resulting system

$$\begin{array}{rcrcl} -a & + & b & = & 0 \\ & & b & = & 1 \\ a & + & b & = & 2 \\ 2a & + & b & = & 1. \end{array}$$

Since

$$A^T A = \begin{bmatrix} -1 & 0 & 1 & 2 \\ 1 & 1 & 1 & 1 \end{bmatrix} \begin{bmatrix} -1 & 1 \\ 0 & 1 \\ 1 & 1 \\ 2 & 1 \end{bmatrix} = \begin{bmatrix} 6 & 2 \\ 2 & 4 \end{bmatrix} \quad \text{and} \quad A^T \mathbf{b} = \begin{bmatrix} -1 & 0 & 1 & 2 \\ 1 & 1 & 1 & 1 \end{bmatrix} \begin{bmatrix} 0 \\ 1 \\ 2 \\ 1 \end{bmatrix} = \begin{bmatrix} 4 \\ 4 \end{bmatrix},$$

the system $A^T A \mathbf{x}^* = A^T \mathbf{b}$ is

$$\begin{array}{rcrcl} 6x_1^* & + & 2x_2^* & = & 4 \\ 2x_1^* & + & 4x_2^* & = & 4. \end{array}$$

The solution is

$$\mathbf{x}^* = \begin{bmatrix} 2/5 \\ 4/5 \end{bmatrix},$$

and thus $f(x) = \frac{2}{5}x + \frac{4}{5}$ is the line which best fits the given data.

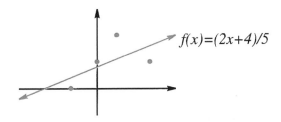

$f(x)=(2x+4)/5$

$\diamond$

Exercises

In Exercises 1 through 7, find the matrix for the orthogonal projection onto the subspace V.

20.1. $V = \text{span}\left(\begin{bmatrix} 2 \\ 1 \end{bmatrix}\right)$

20.2. V is the line $4x_1 + 2x_2 = 0$.

20.3. $V = \text{span}\left(\begin{bmatrix} 2 \\ 1 \\ -3 \end{bmatrix}\right)$

20.4. $V = \text{span}\left(\begin{bmatrix} 2 \\ 1 \\ -3 \end{bmatrix}, \begin{bmatrix} 4 \\ 0 \\ 2 \end{bmatrix}\right)$

20.5. V is the plane $-x_1 + 2x_2 + 3x_3 = 0$.

20.6. $V = \text{span}\left(\begin{bmatrix} 1 \\ 1 \\ 0 \end{bmatrix}, \begin{bmatrix} 3 \\ 1 \\ 2 \end{bmatrix}, \begin{bmatrix} -2 \\ 0 \\ 1 \end{bmatrix}\right)$

20.7. $V = \text{span}\left(\begin{bmatrix} -4 \\ 2 \\ 0 \\ 5 \end{bmatrix}\right)$

In Exercises 8 and 9, find the distance between the point in $\mathbf{R}^3$ and the plane in $\mathbf{R}^3$.

20.8. $(1, 0, 1)$ and $3x_1 - 2x_2 + x_3 = 0$.

20.9. $(1, 2, 3)$ and $x_1 - x_3 = 0$.

In Exercises 10 through 12, find the least squares solution of the system $A\mathbf{x} = \mathbf{b}$.

20.10. $A = \begin{bmatrix} 1 & 2 \\ -2 & 0 \\ 3 & 1 \end{bmatrix}$ $\quad \mathbf{b} = \begin{bmatrix} 1 \\ 4 \\ 0 \end{bmatrix}$

20.11. $A = \begin{bmatrix} 4 & 1 \\ -2 & 5 \\ 1 & 3 \end{bmatrix}$ $\mathbf{b} = \begin{bmatrix} 1 \\ 0 \\ 1 \end{bmatrix}$

20.12. $A = \begin{bmatrix} 1 & 2 \\ 0 & -1 \\ 3 & 1 \\ 2 & 4 \end{bmatrix}$ $\mathbf{b} = \begin{bmatrix} -1 \\ 0 \\ 1 \\ 0 \end{bmatrix}$

20.13. Consider the data points $(1,2), (2,4)$ and $(4,6)$.

  (a) Find the line $f(x) = ax + b$ which best fits the data points.

  (b) Sketch the data points and the linear approximation from part (a).

  (c) Find a quadratic polynomial $f(x) = ax^2 + bx + c$ which contains all three data points.

20.14. Consider the data points $(-2,1)$, $(0,2)$ and $(2,4)$.

  (a) Find the line $f(x) = ax + b$ which best fits the data points.

  (b) Sketch the data points and the linear approximation from part (a).

  (c) Find a quadratic polynomial $f(x) = ax^2 + bx + c$ which contains all three data points.

20.15. Find a quadratic polynomial $f(x) = ax^2 + bx + c$ which best fits the data points $(-1,1)$, $(0,4)$, $(1,3)$ and $(2,5)$.

20.16. Let $P : \mathbf{R}^n \to \mathbf{R}^n$ be the matrix associated with orthogonal projection onto a subspace V of $\mathbf{R}^n$. Show that $P^2 = I_n$.

20.17. We say a matrix A is **symmetric** if $A = A^T$. Show that the matrix P associated with orthogonal projection onto a subspace V of $\mathbf{R}^n$ is a symmetric matrix.

20.18. Show that if A is invertible, then the least squares solution $\mathbf{x}^*$ of $A\mathbf{x} = \mathbf{b}$ is the same as the actual solution of $A\mathbf{x} = \mathbf{b}$.

In Exercises 19 through 22, find the solution of $A\mathbf{x} = \mathbf{b}$ which has the smallest magnitude. That is, find the unique solution in $C(A^T)$.

20.19. $A = \begin{bmatrix} 1 & 2 & 3 \\ 2 & -1 & 1 \end{bmatrix}$, $\mathbf{b} = \begin{bmatrix} 6 \\ 2 \end{bmatrix}$

20.20. $A = \begin{bmatrix} 1 & 2 & 3 \\ 4 & 5 & 6 \\ 7 & 8 & 9 \end{bmatrix}$, $\mathbf{b} = \begin{bmatrix} 1 \\ 4 \\ 7 \end{bmatrix}$

20.21. $A = \begin{bmatrix} 0 & 1 & 1 \\ 1 & 2 & 1 \\ 1 & 1 & 0 \end{bmatrix}$, $\mathbf{b} = \begin{bmatrix} 2 \\ 3 \\ 1 \end{bmatrix}$

20.22. $A = \begin{bmatrix} 0 & 0 & 1 & 1 \\ 1 & 1 & 0 & 0 \\ 0 & 1 & 1 & 0 \end{bmatrix}$, $\mathbf{b} = \begin{bmatrix} 5 \\ 1 \\ 3 \end{bmatrix}$

21 Systems of Coordinates

Each basis for a subspace V of $\mathbf{R}^n$ determines a different **coordinate system** on V. Suppose $\mathcal{B} = \{\mathbf{v}_1, \mathbf{v}_2, \dots, \mathbf{v}_k\}$ is a basis for V, and recall that by Proposition 11.1 any vector $\mathbf{v}$ in V can be written uniquely as a linear combination

$$\mathbf{v} = c_1\mathbf{v}_1 + c_2\mathbf{v}_2 + \cdots + c_n\mathbf{v}_k$$

of the basis vectors. We call the coefficients $c_1, c_2, \dots, c_k$ the **coordinates** of $\mathbf{v}$ with respect to the basis $\mathcal{B}$ and write

$$[\mathbf{v}]_\mathcal{B} = \begin{bmatrix} c_1 \\ c_2 \\ \vdots \\ c_k \end{bmatrix}.$$

Notice that there are k components to this vector, each corresponding to the coefficient of one of the basis vectors.

Example 21.1. Let

$$\mathbf{v}_1 = \begin{bmatrix} 1 \\ 2 \\ 3 \end{bmatrix} \quad \text{and} \quad \mathbf{v}_2 = \begin{bmatrix} 1 \\ 0 \\ 1 \end{bmatrix},$$

and define $V = \text{span}(\mathbf{v}_1, \mathbf{v}_2)$. Then $\mathcal{B} = \{\mathbf{v}_1, \mathbf{v}_2\}$ is a basis for V. The vector

$$\mathbf{v} = 2\mathbf{v}_1 - 3\mathbf{v}_2 = \begin{bmatrix} -1 \\ 4 \\ 3 \end{bmatrix}$$

is a linear combination of $\mathbf{v}_1$ and $\mathbf{v}_2$ and is therefore in V. Its coordinates with respect to $\mathcal{B}$ are

$$[\mathbf{v}]_\mathcal{B} = \begin{bmatrix} 2 \\ -3 \end{bmatrix}.$$

Notice that, although $\mathbf{v}$ is a vector in $\mathbf{R}^3$, it has two coordinates with respect to the basis $\mathcal{B}$ for the two-dimensional subspace V and is thus represented by a vector with two components. $\diamond$

Example 21.2. Let $S = \{e_1, e_2, \dots, e_n\}$ be the standard basis for $\mathbf{R}^n$. Given any vector $\mathbf{x}$ in $\mathbf{R}^n$, we have

$$\mathbf{x} = \begin{bmatrix} x_1 \\ x_2 \\ \vdots \\ x_n \end{bmatrix} = x_1 e_1 + x_2 e_2 + \cdots + x_n e_n$$

so the coordinates of $\mathbf{x}$ with respect to S are

$$[\mathbf{x}]_S = \begin{bmatrix} x_1 \\ x_2 \\ \vdots \\ x_n \end{bmatrix}.$$

Thus our usual representation of a vector in $\mathbf{R}^n$ is in terms of its coordinates with respect to the standard basis. We call these coordinates the **standard coordinates**. $\diamond$

Example 21.3. Let

$$\mathbf{v}_1 = \begin{bmatrix} 2 \\ 1 \end{bmatrix} \qquad \text{and} \qquad \mathbf{v}_2 = \begin{bmatrix} 1 \\ 2 \end{bmatrix}.$$

Then $\mathcal{B} = \{\mathbf{v}_1, \mathbf{v}_2\}$ is a basis for $\mathbf{R}^2$. Given a vector

$$\mathbf{x} = \begin{bmatrix} 4 \\ 5 \end{bmatrix}$$

its $\mathcal{B}$ coordinates are found by solving $\mathbf{x} = c_1 \mathbf{v}_1 + c_2 \mathbf{v}_2$. The solution is $c_1 = 1$ and $c_2 = 2$, so

$$[\mathbf{x}]_\mathcal{B} = \begin{bmatrix} 1 \\ 2 \end{bmatrix}.$$

On the other hand, suppose we are given the coordinates of a vector with respect to $\mathcal{B}$, say

$$[\mathbf{y}]_\mathcal{B} = \begin{bmatrix} -2 \\ 3 \end{bmatrix}.$$

Then, in standard coordinates

$$\mathbf{y} = -2\mathbf{v}_1 + 3\mathbf{v}_2 = \begin{bmatrix} -1 \\ 4 \end{bmatrix}.$$

$\diamond$

We can visualize coordinate systems by drawing the "graph paper" generated by the basis vectors.

Example 21.4. The graph paper generated by the standard basis $S = \{e_1, e_2\}$ for $\mathbf{R}^2$ is shown below.

146

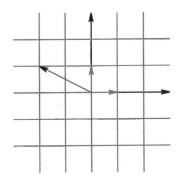

The vector shown,

$$\mathbf{x} = \begin{bmatrix} -2 \\ 1 \end{bmatrix}$$

is obtained by moving from the origin -2 units along the $\mathbf{e}_1$-axis and 1 unit along the $\mathbf{e}_2$-axis.
◇

Example 21.5. The graph paper generated by the basis $\mathcal{B}$ in Example 21.3 is shown below.

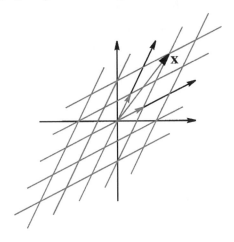

The vector shown is

$$\mathbf{x} = \begin{bmatrix} 4 \\ 5 \end{bmatrix}.$$

From Example 21.3, its $\mathcal{B}$ coordinates are

$$[\mathbf{x}]_{\mathcal{B}} = \begin{bmatrix} 1 \\ 2 \end{bmatrix}.$$

Geometrically $\mathbf{x}$ is obtained by moving from the origin 1 unit along the $\mathbf{v}_1$-axis and 2 units along the $\mathbf{v}_2$-axis.
◇

Change of Basis Matrix

To recover the standard coordinates of a vector from its $\mathcal{B}$ coordinates we use the matrix

$$
C = \begin{bmatrix} | & | & & | \\ \mathbf{v}_1 & \mathbf{v}_2 & \cdots & \mathbf{v}_k \\ | & | & & | \end{bmatrix}
$$

whose columns are the vectors in $\mathcal{B}$. Since

$$
\mathbf{v} = c_1\mathbf{v}_1 + c_2\mathbf{v}_2 + \cdots + c_k\mathbf{v}_k = \begin{bmatrix} | & | & & | \\ \mathbf{v}_1 & \mathbf{v}_2 & \cdots & \mathbf{v}_k \\ | & | & & | \end{bmatrix} \begin{bmatrix} c_1 \\ c_2 \\ \vdots \\ c_k \end{bmatrix}
$$

we have

$$
\boxed{\mathbf{v} = C[\mathbf{v}]_\mathcal{B}.}
$$

Thus the standard coordinates are obtained by multiplying the $\mathcal{B}$ coordinates by C, and we therefore refer to the matrix C as the **change of basis matrix** for the basis $\mathcal{B}$. On the other hand, to find the $\mathcal{B}$ coordinates of a vector $\mathbf{v}$ in V, we need to solve the system $C[\mathbf{v}]_\mathcal{B} = \mathbf{v}$ for $[\mathbf{v}]_\mathcal{B}$. Since the columns of C are basis vectors of V, we are guaranteed that this system has a unique solution for every $\mathbf{v}$ in V.

Example 21.6. Let V be the plane spanned by

$$
\mathbf{v}_1 = \begin{bmatrix} 1 \\ 2 \\ 3 \end{bmatrix} \qquad \text{and} \qquad \mathbf{v}_2 = \begin{bmatrix} 1 \\ 0 \\ 1 \end{bmatrix},
$$

and suppose that the coordinates of some vector $\mathbf{v}$ in V with respect to the basis $\mathcal{B} = \{\mathbf{v}_1, \mathbf{v}_2\}$ are

$$
[\mathbf{v}]_\mathcal{B} = \begin{bmatrix} 7 \\ -4 \end{bmatrix}.
$$

Then since the change of basis matrix is

$$
C = \begin{bmatrix} 1 & 1 \\ 2 & 0 \\ 3 & 1 \end{bmatrix},
$$

the standard coordinates of $\mathbf{v}$ are

$$
\mathbf{v} = C[\mathbf{v}]_\mathcal{B} = \begin{bmatrix} 1 & 1 \\ 2 & 0 \\ 3 & 1 \end{bmatrix} \begin{bmatrix} 7 \\ -4 \end{bmatrix} = \begin{bmatrix} 3 \\ 14 \\ 17 \end{bmatrix}.
$$

On the other hand, it can be seen by taking the cross product of the vectors $\mathbf{v}_1$ and $\mathbf{v}_2$ that the plane has equation $x + y - z = 0$. So, for example, the vector

$$\mathbf{w} = \begin{bmatrix} 8 \\ -6 \\ 2 \end{bmatrix}$$

is in V. What are its $\mathcal{B}$ coordinates? To find out, we must solve $C[\mathbf{w}]_{\mathcal{B}} = \mathbf{w}$ for the components c_1 and c_2 of $[\mathbf{w}]_{\mathcal{B}}$. That is, we must solve

$$\begin{bmatrix} 1 & 1 \\ 2 & 0 \\ 3 & 1 \end{bmatrix} \begin{bmatrix} c_1 \\ c_2 \end{bmatrix} = \begin{bmatrix} 8 \\ -6 \\ 2 \end{bmatrix}$$

for c_1 and c_2. The unique solution is $c_1 = -3$, $c_2 = 11$. Thus

$$[\mathbf{w}]_{\mathcal{B}} = \begin{bmatrix} -3 \\ 11 \end{bmatrix}$$

are the $\mathcal{B}$ coordinates of $\mathbf{w}$. $\diamond$

In the case $V = \mathbf{R}^n$, $\mathcal{B}$ is a basis for $\mathbf{R}^n$, and the matrix C is invertible, and we therefore have the relation

$$\boxed{[\mathbf{v}]_{\mathcal{B}} = C^{-1}\mathbf{v}}$$

so the $\mathcal{B}$ coordinates are obtained by multiplying the standard coordinates by C^{-1}.

Example 21.7. Consider once again the basis from Example 21.3. The change of basis matrix is

$$C = \begin{bmatrix} 2 & 1 \\ 1 & 2 \end{bmatrix}$$

so the $\mathcal{B}$ coordinates of

$$\mathbf{v} = \begin{bmatrix} 4 \\ 5 \end{bmatrix}$$

are

$$[\mathbf{v}]_{\mathcal{B}} = C^{-1}\mathbf{v} = \frac{1}{3} \begin{bmatrix} 2 & -1 \\ -1 & 2 \end{bmatrix} \begin{bmatrix} 4 \\ 5 \end{bmatrix} = \begin{bmatrix} 1 \\ 2 \end{bmatrix}$$

which agrees with our earlier calculation.

$\diamond$

Matrix of a Linear Transformation

One reason for studying bases other than the standard basis is to better understand linear transformations. Let $\mathbf{T} : \mathbf{R}^n \to \mathbf{R}^n$ be a linear transformation. Recall that the matrix A for $\mathbf{T}$ is the matrix such that

$$\mathbf{T}(\mathbf{v}) = A\mathbf{v}$$

for all $\mathbf{v}$ in $\mathbf{R}^n$. Now let $\mathcal{B} = \{\mathbf{v}_1, \ldots, \mathbf{v}_n\}$ be a basis for $\mathbf{R}^n$. We say that B is the matrix for $\mathbf{T}$ with respect to $\mathcal{B}$ if

$$[\mathbf{T}(\mathbf{v})]_\mathcal{B} = B[\mathbf{v}]_\mathcal{B}$$

for all $\mathbf{v}$ in $\mathbf{R}^n$. That is, B is the matrix which sends the $\mathcal{B}$-coordinates of $\mathbf{v}$ to the $\mathcal{B}$-coordinates of $\mathbf{T}(\mathbf{v})$. The matrix A is simply the matrix for $\mathbf{T}$ with respect to the standard basis. The relationship between A and B can be found in terms of the change of basis matrix C for $\mathcal{B}$. Using the definition above we have

$$B[\mathbf{v}]_\mathcal{B} = [\mathbf{T}(\mathbf{v})]_\mathcal{B} = C^{-1}\mathbf{T}(\mathbf{v}) = C^{-1}A\mathbf{v} = C^{-1}AC[\mathbf{v}]_\mathcal{B}.$$

Thus

$$\boxed{B = C^{-1}AC.}$$

This relationship is illustrated in the following diagram.

$$
\begin{array}{ccc}
\mathbf{v} & \xrightarrow{\;\;A\;\;} & \mathbf{T}(\mathbf{v}) & \longleftarrow & \text{standard coordinates} \\
{\scriptstyle C}\Big\uparrow & & \Big\downarrow{\scriptstyle C^{-1}} & & \\
[\mathbf{v}]_\mathcal{B} & \xrightarrow[\;\;B\;\;]{} & [\mathbf{T}(\mathbf{v})]_\mathcal{B} & \longleftarrow & \mathcal{B} \text{ coordinates}
\end{array}
$$

In the diagram, both rows represent the action of the linear transformation $\mathbf{T}$ on some vector $\mathbf{v}$ in $\mathbf{R}^n$. The top row states that this action is performed in standard coordinates by multiplying by the matrix A, while the bottom row states that it is performed in $\mathcal{B}$ coordinates by multiplying by the matrix B. So both A and B represent the same linear transformation $\mathbf{T}$, but in different systems of coordinates.

The relation $B = C^{-1}AC$ can be understood as two different ways of obtaining the $\mathcal{B}$-coordinates of $\mathbf{T}(\mathbf{v})$ from the $\mathcal{B}$-coordinates of $\mathbf{v}$. Beginning from $[\mathbf{v}]_\mathcal{B}$ in the lower left corner, to obtain $[\mathbf{T}(\mathbf{v})]_\mathcal{B}$ we can either multiply by B, or multiply first by C, then by A and then by C^{-1}, i.e. multiply by $C^{-1}AC$. The latter method corresponds to first changing to standard coordinates, computing $\mathbf{T}(\mathbf{v})$ in standard coordinates, and then changing back to $\mathcal{B}$ coordinates.

The equation $B = C^{-1}AC$ may be solved for A by multiplying on the left by C and on the right by C^{-1}. This results in

$$A = CBC^{-1}.$$

The purpose of considering coordinate systems other than the standard coordinates is that for an appropriately chosen basis $\mathcal{B}$, the matrix for a transformation with respect to $\mathcal{B}$ is very simple, and thus provides greater insight into the nature of the transformation.

Example 21.8. Let $\mathbf{T} : \mathbf{R}^2 \to \mathbf{R}^2$ be the transformation with matrix

$$A = \begin{bmatrix} 3 & -2 \\ 2 & -2 \end{bmatrix}$$

with respect to the standard basis and let

$$\mathcal{B} = \left\{ \begin{bmatrix} 2 \\ 1 \end{bmatrix}, \begin{bmatrix} 1 \\ 2 \end{bmatrix} \right\}.$$

The matrix for $\mathbf{T}$ with respect to $\mathcal{B}$ is

$$B = C^{-1}AC = \frac{1}{3} \begin{bmatrix} 2 & -1 \\ -1 & 2 \end{bmatrix} \begin{bmatrix} 3 & -2 \\ 2 & -2 \end{bmatrix} \begin{bmatrix} 2 & 1 \\ 1 & 2 \end{bmatrix} = \begin{bmatrix} 2 & 0 \\ 0 & -1 \end{bmatrix}.$$

So in $\mathcal{B}$-coordinates, $\mathbf{T}$ is represented by a diagonal matrix. It scales the first coordinate by 2 and the second by -1. For instance, if

$$[\mathbf{x}]_\mathcal{B} = \begin{bmatrix} 1 \\ -2 \end{bmatrix}$$

then

$$[\mathbf{T}(\mathbf{x})]_\mathcal{B} = \begin{bmatrix} 2 \\ 2 \end{bmatrix}.$$

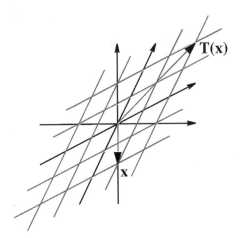

Figure 21.1.

In standard coordinates

$$\mathbf{x} = 1 \begin{bmatrix} 2 \\ 1 \end{bmatrix} - 2 \begin{bmatrix} 1 \\ 2 \end{bmatrix} = \begin{bmatrix} 0 \\ -3 \end{bmatrix}$$

and

$$\mathbf{T}(\mathbf{x}) = 2 \begin{bmatrix} 2 \\ 1 \end{bmatrix} + 2 \begin{bmatrix} 1 \\ 2 \end{bmatrix} = \begin{bmatrix} 6 \\ 6 \end{bmatrix}.$$

It is of course faster to just multiply by A. But suppose we wanted to apply the transformation $\mathbf{T}$ ten times. This would involve multiplying by A ten times, or equivalently, multiplying by A^{10}. Unfortunately computing A^{10} directly is rather tedious. Instead, let's use the relation $A = CBC^{-1}$.

$$A^2 = CBC^{-1}CBC^{-1} = CB^2C^{-1}$$
$$A^3 = A^2A = CB^2C^{-1}CBC^{-1} = CB^3C^{-1}$$
$$\vdots$$
$$A^{10} = CB^{10}C^{-1}$$

The point is that powers of B are very easy to compute.

$$B^2 = \begin{bmatrix} 4 & 0 \\ 0 & 1 \end{bmatrix} \qquad B^3 = \begin{bmatrix} 8 & 0 \\ 0 & -1 \end{bmatrix} \qquad \cdots \qquad B^{10} = \begin{bmatrix} 2^{10} & 0 \\ 0 & (-1)^{10} \end{bmatrix}$$

Thus

$$A^{10} = \begin{bmatrix} 2 & 1 \\ 1 & 2 \end{bmatrix} \begin{bmatrix} 1024 & 0 \\ 0 & 1 \end{bmatrix} \frac{1}{3} \begin{bmatrix} 2 & -1 \\ -1 & 2 \end{bmatrix} = \frac{1}{3} \begin{bmatrix} 2 & 1 \\ 1 & 2 \end{bmatrix} \begin{bmatrix} 2048 & -1024 \\ -1 & 2 \end{bmatrix} = \begin{bmatrix} 1365 & -682 \\ 682 & -340 \end{bmatrix}.$$

$\diamond$

Now suppose that we do not know the matrix for a linear transformation $\mathbf{T}$. We can find its matrix B with respect to a basis $\mathcal{B}$ by evaluating $\mathbf{T}$ on the basis vectors $\{\mathbf{v}_1, \mathbf{v}_2, \ldots, \mathbf{v}_n\}$ in $\mathcal{B}$. By definition $[\mathbf{T}(\mathbf{v})]_{\mathcal{B}} = B[\mathbf{v}]_{\mathcal{B}}$ for all $\mathbf{v}$ in $\mathbf{R}^n$. Observe that

$$\mathbf{v}_1 = 1\mathbf{v}_1 + 0\mathbf{v}_2 + \cdots + 0\mathbf{v}_n$$

so

$$[\mathbf{v}_1]_{\mathcal{B}} = \begin{bmatrix} 1 \\ 0 \\ \vdots \\ 0 \end{bmatrix} = \mathbf{e}_1,$$

and likewise $[\mathbf{v}_j]_{\mathcal{B}} = \mathbf{e}_j$ for each j. Thus

$$[\mathbf{T}(\mathbf{v}_j)]_{\mathcal{B}} = B\mathbf{e}_j$$

is the jth column of B. So

$$B = \left[\begin{array}{cccc} \vphantom{T} & \vphantom{T} & & \vphantom{T} \\ \left[\mathbf{T}(\mathbf{v}_1)\right]_{\mathcal{B}} & \left[\mathbf{T}(\mathbf{v}_2)\right]_{\mathcal{B}} & \cdots & \left[\mathbf{T}(\mathbf{v}_n)\right]_{\mathcal{B}} \\ & & & \end{array}\right]$$

is the matrix for $\mathbf{T}$ with respect to $\mathcal{B}$. Knowing the matrix B then allows us to find the matrix for $\mathbf{T}$ in standard coordinates via the relation $A = CBC^{-1}$.

Example 21.9. Let $\mathbf{T}$ be the linear transformation defined by reflection across the line $y = 3x$ in $\mathbf{R}^2$. A nice way to express this transformation is in terms of the vector

$$\mathbf{v}_1 = \begin{bmatrix} 1 \\ 3 \end{bmatrix}$$

which spans the line $y = 3x$ and the vector

$$\mathbf{v}_2 = \begin{bmatrix} -3 \\ 1 \end{bmatrix}$$

which is perpendicular to the line. It is clear geometrically that $\mathbf{T}(\mathbf{v}_1) = \mathbf{v}_1$ and $\mathbf{T}(\mathbf{v}_2) = -\mathbf{v}_2$.

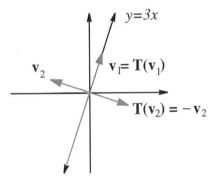

In terms of the basis $\mathcal{B} = \{\mathbf{v}_1, \mathbf{v}_2\}$

$$\begin{array}{rcccc} \mathbf{T}(\mathbf{v}_1) & = & 1\mathbf{v}_1 & + & 0\mathbf{v}_2 \\ \mathbf{T}(\mathbf{v}_2) & = & 0\mathbf{v}_1 & + & (-1)\mathbf{v}_2, \end{array}$$

and therefore

$$[\mathbf{T}(\mathbf{v}_1)]_{\mathcal{B}} = \begin{bmatrix} 1 \\ 0 \end{bmatrix} \quad \text{and} \quad [\mathbf{T}(\mathbf{v}_2)]_{\mathcal{B}} = \begin{bmatrix} 0 \\ -1 \end{bmatrix}.$$

So the matrix for $\mathbf{T}$ with respect to $\mathcal{B}$ is

$$B = \begin{bmatrix} 1 & 0 \\ 0 & -1 \end{bmatrix}.$$

In standard coordinates, the matrix for $\mathbf{T}$ is

$$A = CBC^{-1} = \begin{bmatrix} 1 & -3 \\ 3 & 1 \end{bmatrix} \begin{bmatrix} 1 & 0 \\ 0 & -1 \end{bmatrix} \frac{1}{10} \begin{bmatrix} 1 & 3 \\ -3 & 1 \end{bmatrix} = \begin{bmatrix} -\frac{4}{5} & \frac{3}{5} \\ \frac{3}{5} & \frac{4}{5} \end{bmatrix}.$$

$\diamondsuit$

Example 21.10. Let V be the plane $x_1 + 2x_2 + 5x_3 = 0$ in $\mathbf{R}^3$. Find the matrix for $\mathbf{Proj}_V$, the orthogonal projection onto V.

Solution. Let

$$\mathbf{v}_1 = \begin{bmatrix} -2 \\ 1 \\ 0 \end{bmatrix}, \qquad \mathbf{v}_2 = \begin{bmatrix} -5 \\ 0 \\ 1 \end{bmatrix} \qquad \text{and} \qquad \mathbf{v}_3 = \begin{bmatrix} 1 \\ 2 \\ 5 \end{bmatrix}.$$

Then $\{\mathbf{v}_1, \mathbf{v}_2\}$ is a basis for V and $\{\mathbf{v}_3\}$ is a basis for $V^\perp$, so $\mathcal{B} = \{\mathbf{v}_1, \mathbf{v}_2, \mathbf{v}_3\}$ is a basis for $\mathbf{R}^3$. Since

$$\mathbf{Proj}_V(\mathbf{v}_1) = \mathbf{v}_1 = 1\mathbf{v}_1 + 0\mathbf{v}_2 + 0\mathbf{v}_3$$
$$\mathbf{Proj}_V(\mathbf{v}_2) = \mathbf{v}_2 = 0\mathbf{v}_1 + 1\mathbf{v}_2 + 0\mathbf{v}_3$$
$$\mathbf{Proj}_V(\mathbf{v}_3) = \mathbf{0} = 0\mathbf{v}_1 + 0\mathbf{v}_2 + 0\mathbf{v}_3$$

the matrix for $\mathbf{Proj}_V$ with respect to $\mathcal{B}$ is

$$B = \begin{bmatrix} 1 & 0 & 0 \\ 0 & 1 & 0 \\ 0 & 0 & 0 \end{bmatrix}.$$

The change of basis matrix and its inverse are

$$C = \begin{bmatrix} -2 & -5 & 1 \\ 1 & 0 & 2 \\ 0 & 1 & 5 \end{bmatrix} \qquad \text{and} \qquad C^{-1} = \frac{1}{30} \begin{bmatrix} -2 & 26 & -10 \\ -5 & -10 & 5 \\ 1 & 2 & 5 \end{bmatrix}$$

so the matrix for $\mathbf{Proj}_V$ in standard coordinates is

$$A = CBC^{-1} = \frac{1}{30} \begin{bmatrix} 29 & -2 & -5 \\ -2 & 26 & -10 \\ -5 & -10 & 5 \end{bmatrix}.$$

$\diamond$

Similar Matrices

Motivated by the preceding discussion, we make the following definition. Let A and B be $n \times n$ matrices. We say A is **similar** to B if

$$\boxed{A = CBC^{-1}}$$

for some matrix C. As mentioned earlier, this relation means that A and B represent the same linear transformation, but in different coordinates. Similar matrices have the following properties, which we leave as exercises.

Proposition 21.1.

1. Any $n \times n$ matrix A is similar to itself.

2. If A is similar to B then B is similar to A.

3. If A is similar to B and B is similar to C, then A is similar to C.

4. If A is similar to B then $\det(A) = \det(B)$.

5. If A is similar to B and A is invertible, then B is invertible and A^{-1} is similar to B^{-1}.

6. If A is similar to B, then A^k is similar to B^k for any positive integer k.

Linear Transformations on Subspaces

Given a subspace V of $\mathbf{R}^n$, we say that a function $\mathbf{T} : V \to V$ is a linear transformation on V if

1. $\mathbf{T}(\mathbf{x} + \mathbf{y}) = \mathbf{T}(\mathbf{x}) + \mathbf{T}(\mathbf{y})$

2. $\mathbf{T}(c\mathbf{x}) = c\mathbf{T}(\mathbf{x})$

for all $\mathbf{x}$ and $\mathbf{y}$ in V and all scalars c in $\mathbf{R}$. Note that the transformation need not be defined on all of $\mathbf{R}^n$. In the same way as we did above, we define the matrix for $\mathbf{T}$ with respect to a basis $\mathcal{B} = \{\mathbf{v}_1, \dots, \mathbf{v}_k\}$ of V to be the matrix B such that

$$[\mathbf{T}(\mathbf{v})]_{\mathcal{B}} = B[\mathbf{v}]_{\mathcal{B}}$$

for all $\mathbf{v}$ in V. By the same reasoning as above, it follows that

$$B = \left[\begin{array}{cccc} \Big[\mathbf{T}(\mathbf{v}_1)\Big]_{\mathcal{B}} & \Big[\mathbf{T}(\mathbf{v}_2)\Big]_{\mathcal{B}} & \cdots & \Big[\mathbf{T}(\mathbf{v}_k)\Big]_{\mathcal{B}} \end{array} \right].$$

Notice that B is a $k \times k$ matrix.

Example 21.11. Let

$$\mathbf{v}_1 = \begin{bmatrix} 2 \\ 1 \\ 2 \end{bmatrix} \qquad \text{and} \qquad \mathbf{v}_2 = \begin{bmatrix} -2 \\ 2 \\ 1 \end{bmatrix}$$

155

and let $V = \text{span}(\mathbf{v}_1, \mathbf{v}_2)$. Define $\mathbf{T} : V \to V$ to be the transformation which reflects vectors in V across the line spanned by $\mathbf{v}_1$. Then $\mathbf{T}(\mathbf{v}_1) = \mathbf{v}_1$, and since $\mathbf{v}_2$ is orthogonal to $\mathbf{v}_1$, $\mathbf{T}(\mathbf{v}_2) = -\mathbf{v}_2$. So in terms of the basis $\mathcal{B} = \{\mathbf{v}_1, \mathbf{v}_2\}$,

$$[\mathbf{T}(\mathbf{v}_1)]_{\mathcal{B}} = \begin{bmatrix} 1 \\ 0 \end{bmatrix} \quad \text{and} \quad [\mathbf{T}(\mathbf{v}_2)]_{\mathcal{B}} = \begin{bmatrix} 0 \\ -1 \end{bmatrix}$$

and thus the matrix for $\mathbf{T}$ with respect to $\mathcal{B}$ is

$$B = \begin{bmatrix} 1 & 0 \\ 0 & -1 \end{bmatrix}.$$

$\Diamond$

Exercises

21.1. Let

$$\mathcal{B} = \left\{ \begin{bmatrix} 3 \\ 1 \end{bmatrix}, \begin{bmatrix} 1 \\ 2 \end{bmatrix} \right\}$$

(a) Write each vector $\mathbf{v}$ in standard coordinates.

   i. $[\mathbf{v}]_{\mathcal{B}} = \begin{bmatrix} 2 \\ -1 \end{bmatrix}$

   ii. $[\mathbf{v}]_{\mathcal{B}} = \begin{bmatrix} 1 \\ 2 \end{bmatrix}$

(b) Find $[\mathbf{v}]_{\mathcal{B}}$ for each vector $\mathbf{v}$.

   i. $\mathbf{v} = \begin{bmatrix} 2 \\ -1 \end{bmatrix}$

   ii. $\mathbf{v} = \begin{bmatrix} 9 \\ 8 \end{bmatrix}$

21.2. Let

$$\mathcal{B} = \left\{ \begin{bmatrix} -4 \\ 1 \end{bmatrix}, \begin{bmatrix} 3 \\ 5 \end{bmatrix} \right\}$$

(a) Write the following vectors in standard coordinates.

   i. $[\mathbf{v}]_{\mathcal{B}} = \begin{bmatrix} 2 \\ 4 \end{bmatrix}$

   ii. $[\mathbf{v}]_{\mathcal{B}} = \begin{bmatrix} 1 \\ 3 \end{bmatrix}$

(b) Write the following vectors in $\mathcal{B}$ coordinates.

   i. $\mathbf{v} = \begin{bmatrix} 2 \\ 4 \end{bmatrix}$

ii. $\mathbf{v} = \begin{bmatrix} 1 \\ 3 \end{bmatrix}$

21.3. Let

$$\mathcal{B} = \left\{ \begin{bmatrix} 1 \\ 0 \\ -1 \end{bmatrix}, \begin{bmatrix} 0 \\ 1 \\ 1 \end{bmatrix}, \begin{bmatrix} 1 \\ 1 \\ -1 \end{bmatrix} \right\}$$

(a) Write each vector $\mathbf{v}$ in standard coordinates.

   i. $[\mathbf{v}]_\mathcal{B} = \begin{bmatrix} 0 \\ 1 \\ 1 \end{bmatrix}$

   ii. $[\mathbf{v}]_\mathcal{B} = \begin{bmatrix} 2 \\ 3 \\ -1 \end{bmatrix}$

(b) Find $[\mathbf{v}]_\mathcal{B}$ for each vector $\mathbf{v}$.

   i. $\mathbf{v} = \begin{bmatrix} 0 \\ 1 \\ 1 \end{bmatrix}$

   ii. $\mathbf{v} = \begin{bmatrix} 2 \\ 2 \\ -1 \end{bmatrix}$

21.4. Let

$$\mathcal{B} = \left\{ \begin{bmatrix} 1 \\ 0 \\ -1 \end{bmatrix}, \begin{bmatrix} 0 \\ 1 \\ 1 \end{bmatrix} \right\}$$

(a) Write each vector $\mathbf{v}$ in standard coordinates.

   i. $[\mathbf{v}]_\mathcal{B} = \begin{bmatrix} 3 \\ 5 \end{bmatrix}$

   ii. $[\mathbf{v}]_\mathcal{B} = \begin{bmatrix} -4 \\ 2 \end{bmatrix}$

(b) Find $[\mathbf{v}]_\mathcal{B}$ for each vector $\mathbf{v}$.

   i. $\mathbf{v} = \begin{bmatrix} 2 \\ -4 \\ -6 \end{bmatrix}$

   ii. $\mathbf{v} = \begin{bmatrix} 1 \\ 5 \\ 4 \end{bmatrix}$

21.5. Let

$$\mathcal{B} = \left\{ \begin{bmatrix} 2 \\ 1 \\ 0 \end{bmatrix}, \begin{bmatrix} 1 \\ 0 \\ 3 \end{bmatrix}, \begin{bmatrix} -1 \\ 2 \\ 3 \end{bmatrix} \right\}$$

(a) Write each vector $\mathbf{v}$ in standard coordinates.

    i. $[\mathbf{v}]_\mathcal{B} = \begin{bmatrix} 4 \\ 3 \\ 1 \end{bmatrix}$

    ii. $[\mathbf{v}]_\mathcal{B} = \begin{bmatrix} 1 \\ 2 \\ 5 \end{bmatrix}$

(b) Find $[\mathbf{v}]_\mathcal{B}$ for each vector $\mathbf{v}$.

    i. $\mathbf{v} = \begin{bmatrix} 0 \\ 3 \\ 4 \end{bmatrix}$

    ii. $\mathbf{v} = \begin{bmatrix} 1 \\ -2 \\ 3 \end{bmatrix}$

21.6. Let

$$\mathcal{B} = \left\{ \begin{bmatrix} 3 \\ 1 \\ -4 \\ 2 \end{bmatrix}, \begin{bmatrix} -2 \\ 5 \\ 1 \\ 1 \end{bmatrix} \right\}.$$

(a) Write each vector $\mathbf{v}$ in standard coordinates.

    i. $[\mathbf{v}]_\mathcal{B} = \begin{bmatrix} 2 \\ -1 \end{bmatrix}$

    ii. $[\mathbf{v}]_\mathcal{B} = \begin{bmatrix} 1 \\ 2 \end{bmatrix}$

(b) Find $[\mathbf{v}]_\mathcal{B}$ for each vector $\mathbf{v}$.

    i. $\mathbf{v} = \begin{bmatrix} 8 \\ -3 \\ -9 \\ 3 \end{bmatrix}$

    ii. $\mathbf{v} = \begin{bmatrix} -12 \\ 13 \\ 11 \\ -1 \end{bmatrix}$

21.7. Let

$$\mathcal{B} = \left\{ \begin{bmatrix} 1 \\ 0 \\ 2 \\ -1 \end{bmatrix}, \begin{bmatrix} 0 \\ 2 \\ 1 \\ 1 \end{bmatrix}, \begin{bmatrix} 2 \\ 1 \\ -1 \\ 0 \end{bmatrix} \right\}.$$

(a) Write each vector $\mathbf{v}$ in standard coordinates.

    i. $[\mathbf{v}]_\mathcal{B} = \begin{bmatrix} 3 \\ -1 \\ 0 \end{bmatrix}$

    ii. $[\mathbf{v}]_\mathcal{B} = \begin{bmatrix} 4 \\ -2 \\ 1 \end{bmatrix}$

(b) Find $[\mathbf{v}]_\mathcal{B}$ for each vector $\mathbf{v}$.

    i. $\mathbf{v} = \begin{bmatrix} -4 \\ 5 \\ 0 \\ 5 \end{bmatrix}$

    ii. $\mathbf{v} = \begin{bmatrix} 8 \\ -5 \\ -3 \\ -6 \end{bmatrix}$

21.8. Let $\mathcal{B}$ be the basis $\{\mathbf{v}_1, \mathbf{v}_2, \mathbf{v}_3\}$ of $\mathbf{R}^3$ where

$$\mathbf{v}_1 = \begin{bmatrix} 1 \\ 1 \\ 1 \end{bmatrix} \qquad \mathbf{v}_2 = \begin{bmatrix} 2 \\ 3 \\ 0 \end{bmatrix} \qquad \mathbf{v}_3 = \begin{bmatrix} -1 \\ 2 \\ -6 \end{bmatrix}.$$

(a) Let $\mathbf{T} : \mathbf{R}^3 \to \mathbf{R}^3$ be the linear transformation which satisfies

$$\mathbf{T}(\mathbf{v}_1) = \mathbf{v}_2 \qquad \mathbf{T}(\mathbf{v}_2) = \mathbf{v}_3 \qquad \mathbf{T}(\mathbf{v}_3) = \mathbf{v}_1.$$

Write down the matrix B for $\mathbf{T}$ with respect to the basis $\mathcal{B}$ and use this to find the matrix A for $\mathbf{T}$ with respect to the standard basis.

(b) Compute B^3 and use this to calculate A^3.

(c) Use the result in part (b) to find A^{2000}.

21.9. Let $\mathcal{B}$ be the basis $\{\mathbf{v}_1, \mathbf{v}_2, \mathbf{v}_3\}$ of $\mathbf{R}^3$ where

$$\mathbf{v}_1 = \begin{bmatrix} 2 \\ 1 \\ 0 \end{bmatrix} \qquad \mathbf{v}_2 = \begin{bmatrix} 0 \\ 4 \\ 1 \end{bmatrix} \qquad \mathbf{v}_3 = \begin{bmatrix} -3 \\ 0 \\ 2 \end{bmatrix}.$$

159

Let $\mathbf{T} : \mathbf{R}^3 \to \mathbf{R}^3$ be the linear transformation which satisfies

$$\mathbf{T}(\mathbf{v}_1) = 2\mathbf{v}_1 - 3\mathbf{v}_2 \qquad \mathbf{T}(\mathbf{v}_2) = \mathbf{v}_1 + 2\mathbf{v}_2 - 4\mathbf{v}_3 \qquad \mathbf{T}(\mathbf{v}_3) = -\mathbf{v}_1 + 5\mathbf{v}_3.$$

Write down the matrix B for $\mathbf{T}$ with respect to the basis $\mathcal{B}$ and use this to find the matrix A for $\mathbf{T}$ with respect to the standard basis in $\mathbf{R}^3$.

21.10. Let $\mathbf{Proj}_L : \mathbf{R}^2 \to \mathbf{R}^2$ be projection onto the line L spanned by $\mathbf{v} = \begin{bmatrix} 2 \\ 3 \end{bmatrix}$. Let

$$\mathcal{B} = \left\{ \begin{bmatrix} 2 \\ 3 \end{bmatrix}, \begin{bmatrix} -3 \\ 2 \end{bmatrix} \right\}.$$

Find the matrix B for $\mathbf{Proj}_L$ with respect to $\mathcal{B}$. Use this to find the matrix A for $\mathbf{Proj}_L$ with respect to the standard basis in $\mathbf{R}^2$.

21.11. Let $\mathbf{Ref}_L : \mathbf{R}^2 \to \mathbf{R}^2$ be reflection across the line L spanned by $\mathbf{v} = \begin{bmatrix} 1 \\ -2 \end{bmatrix}$. Let

$$\mathcal{B} = \left\{ \begin{bmatrix} 1 \\ -2 \end{bmatrix}, \begin{bmatrix} 2 \\ 1 \end{bmatrix} \right\}.$$

Find the matrix B for $\mathbf{Ref}_L$ with respect to $\mathcal{B}$. Use this to find the matrix A for $\mathbf{Ref}_L$ with respect to the standard basis in $\mathbf{R}^2$.

21.12. Let L be the line in $\mathbf{R}^3$ spanned by

$$\mathbf{v}_1 = \begin{bmatrix} 1 \\ 1 \\ 1 \end{bmatrix}$$

(a) Find a basis $\{\mathbf{v}_2, \mathbf{v}_3\}$ for the plane perpendicular to L, and verify that $\mathcal{B} = \{\mathbf{v}_1, \mathbf{v}_2, \mathbf{v}_3\}$ is a basis for $\mathbf{R}^3$.

(b) Let $\mathbf{Proj}_L$ denote the projection onto the line L. Find the matrix B for $\mathbf{Proj}_L$ with respect to the basis $\mathcal{B}$.

(c) Use your answer to part (b) to find the matrix A for $\mathbf{Proj}_L$ with respect to the standard basis for $\mathbf{R}^3$.

21.13. Let $\mathbf{v}_1, \mathbf{v}_2$ be two linearly independent vectors in $\mathbf{R}^3$. Let $\mathbf{Proj}_P : \mathbf{R}^3 \to \mathbf{R}^3$ be projection onto the plane P spanned by $\{\mathbf{v}_1, \mathbf{v}_2\}$. Let $\mathbf{v}_3$ be a normal vector to this plane. Let

$$\mathcal{B} = \{\mathbf{v}_1, \mathbf{v}_2, \mathbf{v}_3\}.$$

(a) Find the matrix B for $\mathbf{Proj}_P$ with respect to $\mathcal{B}$.

(b) Let

$$\mathbf{v}_1 = \begin{bmatrix} 1 \\ 0 \\ 2 \end{bmatrix} \qquad \mathbf{v}_2 = \begin{bmatrix} -2 \\ 1 \\ 0 \end{bmatrix} \qquad \mathbf{v}_3 = \begin{bmatrix} 2 \\ 4 \\ -1 \end{bmatrix}.$$

Use your answer to part (a) to find the matrix A for $\mathbf{Proj}_P$ with respect to the standard basis for $\mathbf{R}^3$.

21.14. Let $\mathbf{v}_1, \mathbf{v}_2$ be two linearly independent vectors in $\mathbf{R}^3$. Let $\mathbf{Ref}_P : \mathbf{R}^3 \to \mathbf{R}^3$ be reflection across the plane P spanned by $\{\mathbf{v}_1, \mathbf{v}_2\}$. Let $\mathbf{v}_3$ be a normal vector to this plane. Let

$$\mathcal{B} = \{\mathbf{v}_1, \mathbf{v}_2, \mathbf{v}_3\}.$$

(a) Find the matrix B for $\mathbf{Ref}_P$ with respect to $\mathcal{B}$.

(b) Let

$$\mathbf{v}_1 = \begin{bmatrix} -1 \\ 1 \\ 0 \end{bmatrix} \qquad \mathbf{v}_2 = \begin{bmatrix} 1 \\ 0 \\ 2 \end{bmatrix} \qquad \mathbf{v}_3 = \begin{bmatrix} 2 \\ 2 \\ -1 \end{bmatrix}.$$

Use your answer to part (a) to find the matrix A for $\mathbf{Ref}_P$ with respect to the standard basis for $\mathbf{R}^3$.

21.15. Let $\mathbf{v}_1 = \begin{bmatrix} 1 \\ 3 \\ 2 \end{bmatrix}$. Let L be the line in $\mathbf{R}^3$ spanned by $\mathbf{v}_1$. Let P be the plane in $\mathbf{R}^3$ which satisfies $4x_1 - 2x_2 + x_3 = 0$. Let $\mathbf{T} : P \to P$ be the linear transformation which projects vectors in P onto L.

(a) Find a vector $\mathbf{v}_2$ in P orthogonal to $\mathbf{v}_1$.

(b) Let $\mathcal{B} = \{\mathbf{v}_1, \mathbf{v}_2\}$. Find the matrix for $\mathbf{T}$ with respect to $\mathcal{B}$.

21.16. Prove Proposition 21.1,

    (a) Part 1.

    (b) Part 2.

    (c) Part 3.

    (d) Part 4.

    (e) Part 5.

    (f) Part 6.

21.17. *True/False*

    (a) If A is similar to B, then A^T is similar to B^T.

    (b) If A is similar to B and C is similar to D, then AC is similar to BD.

    (c) Let $\mathbf{T} : \mathbf{R}^n \to \mathbf{R}^n$ be a linear transformation. If B_1 is the matrix for $\mathbf{T}$ with respect to a basis $\mathcal{B}_1$ and B_2 is the matrix for $\mathbf{T}$ with respect to a basis $\mathcal{B}_2$, then B_1 is similar to B_2.

22 Orthonormal Bases

A set $\{\mathbf{v}_1, \ldots, \mathbf{v}_k\}$ consisting of mutually orthogonal unit vectors is called an **orthonormal** set. In terms of dot products, we have

$$\mathbf{v}_i \cdot \mathbf{v}_j = \begin{cases} 0 & i \neq j \\ 1 & i = j \end{cases}$$

for $1 \leq i \leq k$ and $1 \leq j \leq k$. A basis for a subspace V of $\mathbf{R}^n$ consisting of mutually orthogonal unit vectors is called an **orthonormal basis** for V.

Example 22.1. The standard basis $\{\mathbf{e}_1, \ldots, \mathbf{e}_n\}$ is an orthonormal basis for $\mathbf{R}^n$. $\diamond$

Example 22.2. Let

$$\mathbf{v}_1 = \begin{bmatrix} 1/3 \\ 2/3 \\ 2/3 \end{bmatrix} \quad \text{and} \quad \mathbf{v}_2 = \begin{bmatrix} 2/3 \\ 1/3 \\ -2/3 \end{bmatrix}.$$

Since $\mathbf{v}_1 \cdot \mathbf{v}_2 = 0$ and $\mathbf{v}_1 \cdot \mathbf{v}_1 = \mathbf{v}_2 \cdot \mathbf{v}_2 = 1$, the set $\{\mathbf{v}_1, \mathbf{v}_2\}$ is an orthonormal set. It is an orthonormal basis for the plane V that it spans. $\diamond$

We next illustrate some of the advantages of orthonormal bases. The first is that it is very easy to determine the coordinates of any given vector with respect to an orthonormal basis. To see this, suppose $\{\mathbf{v}_1, \ldots, \mathbf{v}_k\}$ is an orthonormal basis for V and let $\mathbf{v}$ be any vector in V. Then

$$\mathbf{v} = c_1 \mathbf{v}_1 + c_2 \mathbf{v}_2 + \cdots + c_k \mathbf{v}_k$$

for some coefficients c_1 through c_k. Taking the dot product of both sides with one of the basis vectors $\mathbf{v}_i$ yields

$$\mathbf{v} \cdot \mathbf{v}_i = c_1 \mathbf{v}_1 \cdot \mathbf{v}_i + c_2 \mathbf{v}_2 \cdot \mathbf{v}_i + \cdots + c_k \mathbf{v}_k \cdot \mathbf{v}_i.$$

All of the dot products on the right hand side vanish, except $\mathbf{v}_i \cdot \mathbf{v}_i$ which equals 1, so the right hand side simplifies to just c_i. Thus $c_i = \mathbf{v} \cdot \mathbf{v}_i$, so the coefficient of $\mathbf{v}_i$ is simply the dot product of $\mathbf{v}$ with $\mathbf{v}_i$. This proves the following result.

Proposition 22.1. Let $\{\mathbf{v}_1, \ldots, \mathbf{v}_k\}$ be an orthonormal basis for V. Then

$$\mathbf{v} = (\mathbf{v} \cdot \mathbf{v}_1)\mathbf{v}_1 + (\mathbf{v} \cdot \mathbf{v}_2)\mathbf{v}_2 + \cdots + (\mathbf{v} \cdot \mathbf{v}_k)\mathbf{v}_k$$

for all $\mathbf{v}$ in V.

Example 22.3. Let

$$\mathbf{v}_1 = \begin{bmatrix} 3/5 \\ 4/5 \end{bmatrix} \qquad \text{and} \qquad \mathbf{v}_2 = \begin{bmatrix} -4/5 \\ 3/5 \end{bmatrix}.$$

Then $\mathcal{B} = \{\mathbf{v}_1, \mathbf{v}_2\}$ is an orthonormal basis for $\mathbf{R}^2$. Let

$$\mathbf{v} = \begin{bmatrix} 7 \\ -4 \end{bmatrix}.$$

Since $\mathbf{v} \cdot \mathbf{v}_1 = 1$ and $\mathbf{v} \cdot \mathbf{v}_2 = -8$, we have $\mathbf{v} = \mathbf{v}_1 - 8\mathbf{v}_2$, so the $\mathcal{B}$ coordinates of $\mathbf{v}$ are

$$[\mathbf{v}]_{\mathcal{B}} = \begin{bmatrix} 1 \\ -8 \end{bmatrix}.$$

$\Diamond$

Another benefit of having an orthonormal basis for V is that the formula for the orthogonal projection onto V is simplified. Recall once again that any vector $\mathbf{x}$ can be expressed uniquely as $\mathbf{v} + \mathbf{w}$, where $\mathbf{v}$ is the projection of $\mathbf{x}$ onto V and $\mathbf{w}$ is in $V^{\perp}$. In terms of the orthonormal basis for V, this implies

$$\mathbf{x} = \underbrace{c_1\mathbf{v}_1 + c_2\mathbf{v}_2 + \cdots + c_k\mathbf{v}_k}_{\mathbf{v} = \mathbf{Proj}_V(\mathbf{x})} + \mathbf{w}.$$

As above, if we take the dot product of both sides with $\mathbf{v}_i$, all terms on the right hand side vanish, except $c_i\mathbf{v}_i \cdot \mathbf{v}_i = c_i$, and thus $c_i = \mathbf{x} \cdot \mathbf{v}_i$. This proves the following.

Proposition 22.2. Let $\{\mathbf{v}_1, \mathbf{v}_2, \ldots, \mathbf{v}_k\}$ be an orthonormal basis for a subspace V of $\mathbf{R}^n$. Then

$$\mathbf{Proj}_V(\mathbf{x}) = (\mathbf{x} \cdot \mathbf{v}_1)\mathbf{v}_1 + (\mathbf{x} \cdot \mathbf{v}_2)\mathbf{v}_2 + \cdots + (\mathbf{x} \cdot \mathbf{v}_k)\mathbf{v}_k$$

for all $\mathbf{x}$ in $\mathbf{R}^n$.

Now let A be the matrix whose columns are the basis vectors. Then

$$A^T A = \begin{bmatrix} \rule{1em}{0.4pt} & \mathbf{v}_1^T & \rule{1em}{0.4pt} \\ \rule{1em}{0.4pt} & \mathbf{v}_2^T & \rule{1em}{0.4pt} \\ & \vdots & \\ \rule{1em}{0.4pt} & \mathbf{v}_k^T & \rule{1em}{0.4pt} \end{bmatrix} \begin{bmatrix} | & | & & | \\ \mathbf{v}_1 & \mathbf{v}_2 & \cdots & \mathbf{v}_k \\ | & | & & | \end{bmatrix} = \begin{bmatrix} 1 & 0 & \cdots & 0 \\ 0 & 1 & \cdots & 0 \\ \vdots & \vdots & \ddots & \vdots \\ 0 & 0 & \cdots & 1 \end{bmatrix} = I_k$$

since the ij^{th} entry of $A^T A$ is the dot product of $\mathbf{v}_i$ with $\mathbf{v}_j$. As a consequence, the formula $A(A^T A)^{-1}A^T$ simplifies to AA^T.

Proposition 22.3. Let A be a matrix whose columns form an orthonormal basis for V. Then AA^T is the matrix for $\mathbf{Proj}_V$, the orthogonal projection onto V.

Example 22.4. Let $\{\mathbf{v}_1, \mathbf{v}_2\}$ be the orthonormal set from Example 22.2. The matrix for the orthogonal projection onto the plane spanned by these vectors is

$$AA^T = \begin{bmatrix} 1/3 & 2/3 \\ 2/3 & 1/3 \\ 2/3 & -2/3 \end{bmatrix} \begin{bmatrix} 1/3 & 2/3 & 2/3 \\ 2/3 & 1/3 & -2/3 \end{bmatrix} = \frac{1}{9} \begin{bmatrix} 5 & 4 & -2 \\ 4 & 5 & 2 \\ -2 & 2 & 8 \end{bmatrix}.$$

$\diamond$

Orthogonal Matrices

Suppose $\mathcal{B} = \{\mathbf{v}_1, \ldots, \mathbf{v}_n\}$ is an orthonormal basis for $\mathbf{R}^n$. Its change of basis matrix

$$C = \begin{bmatrix} | & | & & | \\ \mathbf{v}_1 & \mathbf{v}_2 & \cdots & \mathbf{v}_n \\ | & | & & | \end{bmatrix}$$

has columns which are mutually orthogonal unit vectors. So, just as above, we have

$$C^T C = I_n$$

and since C is a square matrix, this implies that C is invertible and

$$\boxed{C^{-1} = C^T.}$$

A matrix with this property is called an **orthogonal matrix**. This property simplifies many calculations since it is clearly easier in general to take a transpose than an inverse. In particular, suppose the matrices for a linear transformation $\mathbf{T}$ with respect to the standard basis and $\mathcal{B}$ are A and B, respectively. Then the relationship between A and B becomes

$$B = C^T A C$$
$$A = C B C^T.$$

Example 22.5. Let

$$\mathbf{v}_1 = \begin{bmatrix} 2/3 \\ -2/3 \\ 1/3 \end{bmatrix}, \qquad \mathbf{v}_2 = \begin{bmatrix} 2/3 \\ 1/3 \\ -2/3 \end{bmatrix} \qquad \text{and} \qquad \mathbf{v}_3 = \begin{bmatrix} 1/3 \\ 2/3 \\ 2/3 \end{bmatrix}.$$

164

Then $\mathcal{B} = \{\mathbf{v}_1, \mathbf{v}_2, \mathbf{v}_3\}$ is an orthonormal basis for $\mathbf{R}^3$. Let V be the plane spanned by $\mathbf{v}_1$ and $\mathbf{v}_2$, and let $\mathbf{T} : \mathbf{R}^3 \to \mathbf{R}^3$ be the reflection through the plane V, which sends vectors in $\mathbf{R}^3$ to their mirror image on the opposite side of V. Then

$$\mathbf{T}(\mathbf{v}_1) = \mathbf{v}_1 = 1\mathbf{v}_1 + 0\mathbf{v}_2 + 0\mathbf{v}_3$$
$$\mathbf{T}(\mathbf{v}_2) = \mathbf{v}_2 = 0\mathbf{v}_1 + 1\mathbf{v}_2 + 0\mathbf{v}_3$$
$$\mathbf{T}(\mathbf{v}_3) = -\mathbf{v}_3 = 0\mathbf{v}_1 + 0\mathbf{v}_2 - 1\mathbf{v}_3$$

so the matrix for $\mathbf{T}$ with respect to $\mathcal{B}$ is

$$B = \begin{bmatrix} 1 & 0 & 0 \\ 0 & 1 & 0 \\ 0 & 0 & -1 \end{bmatrix}.$$

Therefore

$$A = CBC^T = \begin{bmatrix} 2/3 & 2/3 & 1/3 \\ -2/3 & 1/3 & 2/3 \\ 1/3 & -2/3 & 2/3 \end{bmatrix} \begin{bmatrix} 1 & 0 & 0 \\ 0 & 1 & 0 \\ 0 & 0 & -1 \end{bmatrix} \begin{bmatrix} 2/3 & -2/3 & 1/3 \\ 2/3 & 1/3 & -2/3 \\ 1/3 & 2/3 & 2/3 \end{bmatrix} = \frac{1}{9} \begin{bmatrix} 7 & -4 & -4 \\ -4 & 1 & -8 \\ -4 & -8 & 1 \end{bmatrix}$$

is the matrix for $\mathbf{T}$ in standard coordinates. $\diamond$

One very important property of orthogonal matrices is that they preserve lengths and angles. This fact is a consequence of the following result.

Proposition 22.4. Let C be any $n \times n$ orthogonal matrix. Then

$$C\mathbf{v} \cdot C\mathbf{w} = \mathbf{v} \cdot \mathbf{w}$$

for all $\mathbf{v}$ and $\mathbf{w}$ in $\mathbf{R}^n$.

Proof. By Proposition 18.2 and the fact that $C^T C = I_n$, we have $C\mathbf{v} \cdot C\mathbf{w} = \mathbf{v} \cdot C^T C\mathbf{w} = \mathbf{v} \cdot \mathbf{w}$ for all $\mathbf{v}$ and $\mathbf{w}$ in $\mathbf{R}^n$. $\square$

Now, since

$$\|C\mathbf{v}\|^2 = C\mathbf{v} \cdot C\mathbf{v} = \mathbf{v} \cdot \mathbf{v} = \|\mathbf{v}\|^2$$

we have $\|C\mathbf{v}\| = \|\mathbf{v}\|$, so C preserves length. To see that C preserves angle, recall that $\mathbf{v} \cdot \mathbf{w} = \|\mathbf{v}\|\|\mathbf{w}\| \cos\theta$, where θ is the angle between $\mathbf{v}$ and $\mathbf{w}$. Thus

$$\cos\theta = \frac{\mathbf{v} \cdot \mathbf{w}}{\|\mathbf{v}\|\|\mathbf{w}\|} = \frac{C\mathbf{v} \cdot C\mathbf{w}}{\|C\mathbf{v}\|\|C\mathbf{w}\|} = \cos\phi,$$

where ϕ is the angle between $C\mathbf{v}$ and $C\mathbf{w}$. Hence $\theta = \phi$.

Gram-Schmidt Process

We next outline a procedure for producing an orthonormal basis from any given basis $\{\mathbf{v}_1, \mathbf{v}_2, \ldots, \mathbf{v}_k\}$ for a subspace V. We first define

$$V_1 = \operatorname{span}(\mathbf{v}_1)$$
$$V_2 = \operatorname{span}(\mathbf{v}_1, \mathbf{v}_2)$$
$$\vdots$$
$$V_k = \operatorname{span}(\mathbf{v}_1, \mathbf{v}_2, \ldots, \mathbf{v}_k) = V.$$

Next, let

$$\mathbf{w}_1 = \frac{\mathbf{v}_1}{\|\mathbf{v}_1\|}.$$

It is clear that $\{\mathbf{w}_1\}$ is an orthonormal basis for V_1. If V is a one-dimensional subspace, then $V = V_1$ and we are done. Otherwise the process continues by letting

$$\mathbf{y}_2 = \mathbf{v}_2 - \mathbf{Proj}_{V_1}(\mathbf{v}_2).$$

By definition, the difference of $\mathbf{v}_2$ and its projection onto V_1 is in $V_1^{\perp}$, and is therefore orthogonal to $\mathbf{w}_1$.

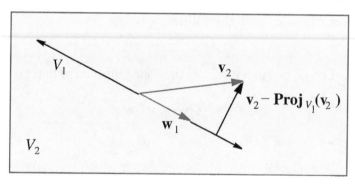

Now since $\mathbf{w}_1$ forms an orthonormal basis for V_1, the projection onto V_1 is given by equation Proposition 22.2. Thus

$$\mathbf{y}_2 = \mathbf{v}_2 - (\mathbf{v}_2 \cdot \mathbf{w}_1)\mathbf{w}_1.$$

By the linear independence of $\mathbf{v}_1$ and $\mathbf{v}_2$, $\mathbf{v}_2$ is not a scalar multiple of $\mathbf{w}_1$, so $\mathbf{y}_2$ is nonzero and we may define

$$\mathbf{w}_2 = \frac{\mathbf{y}_2}{\|\mathbf{y}_2\|}.$$

Then $\{\mathbf{w}_1, \mathbf{w}_2\}$ is an orthonormal basis for V_2. If V is two-dimensional, then $V = V_2$ and we are done. Otherwise we continue as above by letting

$$\mathbf{y}_3 = \mathbf{v}_3 - \mathbf{Proj}_{V_2}(\mathbf{v}_3).$$

166

The vector $\mathbf{y}_3$ is in $V_2^{\perp}$ and is therefore orthogonal to both $\mathbf{w}_1$ and $\mathbf{w}_2$. Since $\{\mathbf{w}_1, \mathbf{w}_2\}$ is an orthonormal basis for V_2, Proposition 22.2 again implies

$$\mathbf{y}_3 = \mathbf{v}_3 - (\mathbf{v}_3 \cdot \mathbf{w}_1)\mathbf{w}_1 - (\mathbf{v}_3 \cdot \mathbf{w}_2)\mathbf{w}_2.$$

By the linear independence of $\{\mathbf{v}_1, \mathbf{v}_2, \mathbf{v}_3\}$, $\mathbf{v}_3$ is not a linear combination of $\mathbf{w}_1$ and $\mathbf{w}_2$, so $\mathbf{y}_3$ is nonzero, and we may define

$$\mathbf{w}_3 = \frac{\mathbf{y}_3}{\|\mathbf{y}_3\|}.$$

Then $\{\mathbf{w}_1, \mathbf{w}_2, \mathbf{w}_3\}$ is an orthonormal basis for V_3. If V is three-dimensional, then $V = V_3$ and we are done. Otherwise the process continues in the same manner until we have constructed an orthonormal basis for $V_k = V$. The process is summarized as follows.

Gram-Schmidt Process

Let $\{\mathbf{v}_1, \mathbf{v}_2, \ldots, \mathbf{v}_k\}$ be a basis for a subspace V of $\mathbf{R}^n$.

1. Let $\mathbf{w}_1 = \mathbf{v}_1/\|\mathbf{v}_1\|$.

2. Having constructed the orthonormal basis $\{\mathbf{w}_1, \ldots, \mathbf{w}_j\}$ for V_j, to construct $\mathbf{w}_{j+1}$ define

$$\begin{aligned}
\mathbf{y}_{j+1} &= \mathbf{v}_{j+1} - \mathbf{Proj}_{V_j}(\mathbf{v}_{j+1}) \\
&= \mathbf{v}_{j+1} - (\mathbf{v}_{j+1} \cdot \mathbf{w}_1)\mathbf{w}_1 - (\mathbf{v}_{j+1} \cdot \mathbf{w}_2)\mathbf{w}_2 - \cdots - (\mathbf{v}_{j+1} \cdot \mathbf{w}_j)\mathbf{w}_j
\end{aligned}$$

and let $\mathbf{w}_{j+1} = \mathbf{y}_{j+1}/\|\mathbf{y}_{j+1}\|$.

Then $\{\mathbf{w}_1, \mathbf{w}_2, \ldots, \mathbf{w}_k\}$ is an orthonormal basis for V.

Example 22.6. Find an orthonormal basis for the plane $x_1 + x_2 + x_3 = 0$.
Solution. First, solving the equation yields the basis vectors

$$\mathbf{v}_1 = \begin{bmatrix} -1 \\ 1 \\ 0 \end{bmatrix} \qquad \text{and} \qquad \mathbf{v}_2 = \begin{bmatrix} -1 \\ 0 \\ 1 \end{bmatrix}$$

for the plane. Applying the Gram-Schmidt process we first have

$$\mathbf{w}_1 = \mathbf{v}_1/\|\mathbf{v}_1\| = \frac{1}{\sqrt{2}} \begin{bmatrix} -1 \\ 1 \\ 0 \end{bmatrix}.$$

Next we let

$$\mathbf{y}_2 = \mathbf{v}_2 - (\mathbf{v}_2 \cdot \mathbf{w}_1)\mathbf{w}_1 = \begin{bmatrix} -1 \\ 0 \\ 1 \end{bmatrix} - \frac{1}{\sqrt{2}}\frac{1}{\sqrt{2}}\begin{bmatrix} -1 \\ 1 \\ 0 \end{bmatrix} = \begin{bmatrix} -1/2 \\ -1/2 \\ 1 \end{bmatrix}$$

and finally

$$\mathbf{w}_2 = \mathbf{y}_2/\|\mathbf{y}_2\| = \frac{1}{\sqrt{6}} \begin{bmatrix} -1 \\ -1 \\ 2 \end{bmatrix}.$$

So

$$\left\{ \frac{1}{\sqrt{2}} \begin{bmatrix} -1 \\ 1 \\ 0 \end{bmatrix}, \frac{1}{\sqrt{6}} \begin{bmatrix} -1 \\ -1 \\ 2 \end{bmatrix} \right\}$$

is an orthonormal basis for the plane. ◇

Example 22.7. Find an orthonormal basis for the subspace

$$V = \operatorname{span} \left(\begin{bmatrix} 1 \\ 1 \\ 0 \\ 0 \end{bmatrix}, \begin{bmatrix} 0 \\ 1 \\ 1 \\ 0 \end{bmatrix}, \begin{bmatrix} 0 \\ 0 \\ 1 \\ 1 \end{bmatrix} \right)$$

of $\mathbf{R}^4$.

Solution. The order in which we write the basis vectors is arbitrary, so to make the first step easier let

$$\mathbf{v}_1 = \begin{bmatrix} 1 \\ 1 \\ 0 \\ 0 \end{bmatrix}, \quad \mathbf{v}_2 = \begin{bmatrix} 0 \\ 0 \\ 1 \\ 1 \end{bmatrix}, \quad \text{and} \quad \mathbf{v}_3 = \begin{bmatrix} 0 \\ 1 \\ 1 \\ 0 \end{bmatrix}.$$

Then

$$\mathbf{w}_1 = \mathbf{v}_1/\|\mathbf{v}_1\| = \frac{1}{\sqrt{2}} \begin{bmatrix} 1 \\ 1 \\ 0 \\ 0 \end{bmatrix}$$

and since $\mathbf{v}_2$ and $\mathbf{w}_1$ are orthogonal,

$$\mathbf{y}_2 = \mathbf{v}_2 - (\mathbf{v}_2 \cdot \mathbf{w}_1)\mathbf{w}_1 = \mathbf{v}_2$$

so

$$\mathbf{w}_2 = \mathbf{y}_2/\|\mathbf{y}_2\| = \frac{1}{\sqrt{2}} \begin{bmatrix} 0 \\ 0 \\ 1 \\ 1 \end{bmatrix}.$$

Finally

$$\mathbf{y}_3 = \mathbf{v}_3 - (\mathbf{v}_3 \cdot \mathbf{w}_1)\mathbf{w}_1 - (\mathbf{v}_3 \cdot \mathbf{w}_2)\mathbf{w}_2 = \begin{bmatrix} 0 \\ 1 \\ 1 \\ 0 \end{bmatrix} - \frac{1}{2}\begin{bmatrix} 1 \\ 1 \\ 0 \\ 0 \end{bmatrix} - \frac{1}{2}\begin{bmatrix} 0 \\ 0 \\ 1 \\ 1 \end{bmatrix} = \begin{bmatrix} -1/2 \\ 1/2 \\ 1/2 \\ -1/2 \end{bmatrix}$$

and thus

$$\mathbf{w}_3 = \mathbf{y}_3/\|\mathbf{y}_3\| = \begin{bmatrix} -1/2 \\ 1/2 \\ 1/2 \\ -1/2 \end{bmatrix}.$$

$\diamond$

As a final example we consider the rather difficult problem of finding the matrix for the rotation about an axis in $\mathbf{R}^3$ which is not one of the coordinate axes.

Example 22.8. Let

$$\mathbf{v}_1 = \begin{bmatrix} 1 \\ 1 \\ 1 \end{bmatrix},$$

and let L be the line spanned by $\mathbf{v}_1$. Define $\mathbf{T}$ to be the rotation through angle θ about the axis L, where the direction of rotation is counterclockwise as viewed from the head of $\mathbf{v}_1$ looking toward the origin. To find the matrix for $\mathbf{T}$ we first construct an appropriate basis for $\mathbf{R}^3$. Vectors in L are fixed by this rotation, so we choose $\mathbf{u}_1 = \mathbf{v}_1/\|\mathbf{v}_1\|$ as the first basis vector. In the plane P perpendicular to L, the rotation behaves like an ordinary rotation in $\mathbf{R}^2$, so we should find a basis for P. The plane P is given by $x_1 + x_2 + x_3 = 0$, so the vectors

$$\mathbf{u}_2 = \frac{1}{\sqrt{2}}\begin{bmatrix} -1 \\ 1 \\ 0 \end{bmatrix} \quad \text{and} \quad \mathbf{u}_3 = \frac{1}{\sqrt{6}}\begin{bmatrix} -1 \\ -1 \\ 2 \end{bmatrix}$$

found in Example 22.6 form an orthonormal basis for P. Thus $\mathcal{B} = \{\mathbf{u}_1, \mathbf{u}_2, \mathbf{u}_3\}$ is an orthonormal basis for $\mathbf{R}^3$. To determine the orientation of $\mathbf{u}_2$ and $\mathbf{u}_3$ in the plane P we take their cross product. Since

$$\mathbf{u}_2 \times \mathbf{u}_3 = \frac{1}{\sqrt{12}}\begin{bmatrix} 2 \\ 2 \\ 2 \end{bmatrix} = +\mathbf{u}_1,$$

looking down on P from the head of $\mathbf{v}_1$ we see the following.

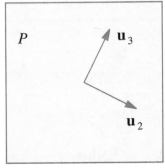

169

So the rotation is such that $\mathbf{u}_2$ is rotated toward $\mathbf{u}_3$. Thus we have

$$\mathbf{T}(\mathbf{u}_1) = \mathbf{u}_1 = 1\mathbf{u}_1 + 0\mathbf{u}_2 + 0\mathbf{u}_3$$
$$\mathbf{T}(\mathbf{u}_2) = (\cos\theta)\mathbf{u}_2 + (\sin\theta)\mathbf{u}_3 = 0\mathbf{u}_1 + (\cos\theta)\mathbf{u}_2 + (\sin\theta)\mathbf{u}_3$$
$$\mathbf{T}(\mathbf{u}_3) = (-\sin\theta)\mathbf{u}_2 + (\cos\theta)\mathbf{u}_3 = 0\mathbf{u}_1 + (-\sin\theta)\mathbf{u}_2 + (\cos\theta)\mathbf{u}_3$$

so the matrix for $\mathbf{T}$ with respect to $\mathcal{B}$ is

$$B = \begin{bmatrix} 1 & 0 & 0 \\ 0 & \cos\theta & -\sin\theta \\ 0 & \sin\theta & \cos\theta \end{bmatrix}.$$

The change of basis matrix for $\mathcal{B}$ is

$$C = \begin{bmatrix} 1/\sqrt{3} & -1/\sqrt{2} & -1/\sqrt{6} \\ 1/\sqrt{3} & 1/\sqrt{2} & -1/\sqrt{6} \\ 1/\sqrt{3} & 0 & 2/\sqrt{6} \end{bmatrix}$$

Since C is an orthogonal matrix, $C^{-1} = C^T$ and the matrix for $\mathbf{T}$ in standard coordinates is

$$A = CBC^T$$
$$= \begin{bmatrix} 1/\sqrt{3} & -1/\sqrt{2} & -1/\sqrt{6} \\ 1/\sqrt{3} & 1/\sqrt{2} & -1/\sqrt{6} \\ 1/\sqrt{3} & 0 & 2/\sqrt{6} \end{bmatrix} \begin{bmatrix} 1 & 0 & 0 \\ 0 & \cos\theta & -\sin\theta \\ 0 & \sin\theta & \cos\theta \end{bmatrix} \begin{bmatrix} 1/\sqrt{3} & 1/\sqrt{3} & 1/\sqrt{3} \\ -1/\sqrt{2} & 1/\sqrt{2} & 0 \\ -1/\sqrt{6} & -1/\sqrt{6} & 2/\sqrt{6} \end{bmatrix}$$

$$= \begin{bmatrix} 1/\sqrt{3} & -1/\sqrt{2} & -1/\sqrt{6} \\ 1/\sqrt{3} & 1/\sqrt{2} & -1/\sqrt{6} \\ 1/\sqrt{3} & 0 & 2/\sqrt{6} \end{bmatrix} \begin{bmatrix} 1/\sqrt{3} & 1/\sqrt{3} & 1/\sqrt{3} \\ -\frac{1}{\sqrt{2}}\cos\theta + \frac{1}{\sqrt{6}}\sin\theta & \frac{1}{\sqrt{2}}\cos\theta + \frac{1}{\sqrt{6}}\sin\theta & -\frac{2}{\sqrt{6}}\sin\theta \\ -\frac{1}{\sqrt{2}}\sin\theta - \frac{1}{\sqrt{6}}\cos\theta & \frac{1}{\sqrt{2}}\sin\theta - \frac{1}{\sqrt{6}}\cos\theta & \frac{2}{\sqrt{6}}\cos\theta \end{bmatrix}$$

$$= \begin{bmatrix} \frac{1}{3} + \frac{2}{3}\cos\theta & \frac{1}{3} - \frac{1}{\sqrt{3}}\sin\theta - \frac{1}{3}\cos\theta & \frac{1}{3} + \frac{1}{\sqrt{3}}\sin\theta - \frac{1}{3}\cos\theta \\ \frac{1}{3} + \frac{1}{\sqrt{3}}\sin\theta - \frac{1}{3}\cos\theta & \frac{1}{3} + \frac{2}{3}\cos\theta & \frac{1}{3} - \frac{1}{\sqrt{3}}\sin\theta - \frac{1}{3}\cos\theta \\ \frac{1}{3} - \frac{1}{\sqrt{3}}\sin\theta - \frac{1}{3}\cos\theta & \frac{1}{3} + \frac{1}{\sqrt{3}}\sin\theta - \frac{1}{3}\cos\theta & \frac{1}{3} + \frac{2}{3}\cos\theta \end{bmatrix}.$$

For instance, the matrix for a $30°$ ($\pi/6$ radian) rotation about L is

$$\begin{bmatrix} \frac{1}{3} + \frac{1}{\sqrt{3}} & \frac{1}{3} - \frac{1}{\sqrt{3}} & \frac{1}{3} \\ \frac{1}{3} & \frac{1}{3} + \frac{1}{\sqrt{3}} & \frac{1}{3} - \frac{1}{\sqrt{3}} \\ \frac{1}{3} - \frac{1}{\sqrt{3}} & \frac{1}{3} & \frac{1}{3} + \frac{1}{\sqrt{3}} \end{bmatrix}.$$

$\diamond$

Exercises

22.1. Let $\mathbf{v}_1 = \begin{bmatrix} 1/\sqrt{5} \\ 2/\sqrt{5} \end{bmatrix}$, $\mathbf{v}_2 = \begin{bmatrix} -2/\sqrt{5} \\ 1/\sqrt{5} \end{bmatrix}$.

(a) Verify that $\mathcal{B} = \{\mathbf{v}_1, \mathbf{v}_2\}$ is an orthonormal basis for $\mathbf{R}^2$.

(b) Write the following vectors in $\mathcal{B}$ coordinates.

   i. $\mathbf{v} = \begin{bmatrix} 1 \\ 3 \end{bmatrix}$

   ii. $\mathbf{v} = \begin{bmatrix} -2 \\ 4 \end{bmatrix}$

22.2. Let $\mathbf{v}_1 = \begin{bmatrix} 1/3 \\ 2/3 \\ 2/3 \end{bmatrix}$, $\mathbf{v}_2 = \begin{bmatrix} 2/3 \\ 1/3 \\ -2/3 \end{bmatrix}$, $\mathbf{v}_3 = \begin{bmatrix} 2/3 \\ -2/3 \\ 1/3 \end{bmatrix}$.

(a) Verify that $\mathcal{B} = \{\mathbf{v}_1, \mathbf{v}_2, \mathbf{v}_3\}$ is an orthonormal basis for $\mathbf{R}^3$.

(b) Write the following vectors in $\mathcal{B}$ coordinates.

   i. $\mathbf{v} = \begin{bmatrix} 0 \\ 1 \\ 0 \end{bmatrix}$

   ii. $\mathbf{v} = \begin{bmatrix} 2 \\ -2 \\ 2 \end{bmatrix}$

22.3. Let $\mathbf{v}_1 = \begin{bmatrix} 1/2 \\ 1/2 \\ 1/2 \\ 1/2 \end{bmatrix}$, $\mathbf{v}_2 = \begin{bmatrix} 1/2 \\ -1/2 \\ -1/2 \\ 1/2 \end{bmatrix}$. Let $V = \mathrm{span}(\mathbf{v}_1, \mathbf{v}_2)$. Let $\mathbf{Proj}_V : \mathbf{R}^4 \to \mathbf{R}^4$ be orthogonal projection onto V. Using the fact that $\{\mathbf{v}_1, \mathbf{v}_2\}$ is an orthonormal basis for V, find the matrix for $\mathbf{Proj}_V$.

22.4. Let $\mathbf{v}_1 = \begin{bmatrix} 1/\sqrt{3} \\ -1/\sqrt{3} \\ 1/\sqrt{3} \end{bmatrix}$, $\mathbf{v}_2 = \begin{bmatrix} 1/\sqrt{5} \\ 2/\sqrt{5} \\ 1/\sqrt{5} \end{bmatrix}$. Let $V = \mathrm{span}(\mathbf{v}_1, \mathbf{v}_2)$. Let $\mathbf{Proj}_V : \mathbf{R}^3 \to \mathbf{R}^3$ be orthogonal projection onto V. Using the fact that $\{\mathbf{v}_1, \mathbf{v}_2\}$ is an orthonormal basis for V, find the matrix for $\mathbf{Proj}_V$.

22.5. Let P be the plane in $\mathbf{R}^3$ which satisfies the equation $-3x_1 + x_2 + 2x_3 = 0$. Let $\mathbf{v}_1 = \begin{bmatrix} 1/\sqrt{3} \\ 1/\sqrt{3} \\ 1/\sqrt{3} \end{bmatrix}$. Notice that $\mathbf{v}_1$ is a unit vector in P.

(a) Find a unit vector $\mathbf{v}_2$ in P perpendicular to $\mathbf{v}_1$.

(b) Notice that $\{\mathbf{v}_1, \mathbf{v}_2\}$ forms an orthonormal basis for P. Find the matrix for $\mathbf{Proj}_P : \mathbf{R}^3 \to \mathbf{R}^3$, the orthogonal projection onto P.

22.6. Let L be the line spanned by $\mathbf{v}_1 = \begin{bmatrix} 1 \\ 2 \\ 2 \end{bmatrix}$, and let $\mathbf{T} : \mathbf{R}^3 \to \mathbf{R}^3$ be the rotation through angle θ about the axis L, where the direction of rotation is counterclockwise as viewed from the head of $\mathbf{v}_1$ looking toward the origin.

(a) Find an orthonormal basis $\{\mathbf{v}_2, \mathbf{v}_3\}$ for the plane $P = L^\perp$.

(b) Find the matrix for $\mathbf{T}$ with respect to the basis $\mathcal{B} = \{\mathbf{v}_1, \mathbf{v}_2, \mathbf{v}_3\}$ for $\mathbf{R}^3$.

(c) Use the answer to part (b) to find the matrix for $\mathbf{T}$ in standard coordinates.

In Exercises 7 through 10 find an orthonormal basis for the subspace V.

22.7. $V = \text{span}\left(\begin{bmatrix} 1 \\ 1 \\ 0 \end{bmatrix}, \begin{bmatrix} 5 \\ -1 \\ 3 \end{bmatrix} \right)$

22.8. $V = \text{span}\left(\begin{bmatrix} -1 \\ -1 \\ 1 \end{bmatrix}, \begin{bmatrix} 2 \\ -2 \\ 3 \end{bmatrix} \right)$

22.9. $V = \text{span}\left(\begin{bmatrix} 1 \\ 1 \\ 1 \\ 0 \end{bmatrix}, \begin{bmatrix} 1 \\ 0 \\ 2 \\ 1 \end{bmatrix}, \begin{bmatrix} 3 \\ -1 \\ 4 \\ 4 \end{bmatrix} \right)$

22.10. $V = \text{span}\left(\begin{bmatrix} 1 \\ 2 \\ 3 \\ 4 \end{bmatrix}, \begin{bmatrix} 1 \\ 1 \\ 0 \\ 0 \end{bmatrix}, \begin{bmatrix} 0 \\ 0 \\ 1 \\ 1 \end{bmatrix} \right)$

23   Eigenvectors

Given a linear transformation $\mathbf{T}$, a *good* basis $\mathcal{B}$ for $\mathbf{R}^n$ is one such that the matrix B for $\mathbf{T}$ with respect to $\mathcal{B}$ is a diagonal matrix. Suppose that such a basis $\mathcal{B} = \{\mathbf{v}_1, \mathbf{v}_2, \ldots, \mathbf{v}_n\}$ exists and that the matrix for $\mathbf{T}$ with respect to $\mathcal{B}$ is

$$B = \begin{bmatrix} \lambda_1 & 0 & \cdots & 0 \\ 0 & \lambda_2 & \cdots & 0 \\ \vdots & \vdots & \ddots & \vdots \\ 0 & 0 & \cdots & \lambda_n \end{bmatrix}.$$

Then

$$[\mathbf{T}(\mathbf{v}_1)]_\mathcal{B} = B[\mathbf{v}_1]_\mathcal{B} = B\mathbf{e}_1 = \begin{bmatrix} \lambda_1 \\ 0 \\ \vdots \\ 0 \end{bmatrix}$$

which means

$$\mathbf{T}(\mathbf{v}_1) = \lambda_1 \mathbf{v}_1.$$

Likewise

$$\mathbf{T}(\mathbf{v}_2) = \lambda_2 \mathbf{v}_2$$

$$\vdots$$

$$\mathbf{T}(\mathbf{v}_n) = \lambda_n \mathbf{v}_n.$$

So when $\mathbf{T}$ is applied to each basis vector, the result is a scalar multiple of that vector. An **eigenvector** of a linear transformation $\mathbf{T} : \mathbf{R}^n \to \mathbf{R}^n$ is a nonzero vector $\mathbf{v}$ such that

$$\boxed{\mathbf{T}(\mathbf{v}) = \lambda \mathbf{v}}$$

for some scalar λ. The number λ is called the **eigenvalue** associated with the eigenvector $\mathbf{v}$. If A is the matrix (in standard coordinates) for $\mathbf{T}$, then $A\mathbf{v} = \lambda\mathbf{v}$, and we say that $\mathbf{v}$ is an eigenvector of A and λ is an eigenvalue of A.

We exclude the zero vector from this definition for two reasons. First, since $\mathbf{T}(\mathbf{0}) = \mathbf{0} = \lambda\mathbf{0}$ for *every* scalar λ, it is unclear what eigenvalue to associate with the zero vector. Also, for the purposes of constructing a basis of $\mathbf{R}^n$, the zero vector is not very useful.

Note however that the number zero could be an eigenvalue. If A is a matrix with nontrivial null space, then any nonzero vector $\mathbf{v}$ in $N(A)$ satisfies $A\mathbf{v} = \mathbf{0} = 0\mathbf{v}$, and is therefore an eigenvector with eigenvalue zero.

Example 23.1. Let

$$A = \begin{bmatrix} 3 & -2 \\ 2 & -2 \end{bmatrix}, \qquad \mathbf{v}_1 = \begin{bmatrix} 2 \\ 1 \end{bmatrix} \qquad \text{and} \qquad \mathbf{v}_2 = \begin{bmatrix} 1 \\ 2 \end{bmatrix}.$$

Then

$$A\mathbf{v}_1 = 2\mathbf{v}_1 \qquad \text{and} \qquad A\mathbf{v}_2 = -\mathbf{v}_2.$$

Thus $\mathbf{v}_1$ is an eigenvector with eigenvalue 2 and $\mathbf{v}_2$ is an eigenvector with eigenvalue -1.

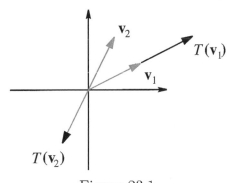

Figure 23.1.

This explains the choice of basis in Example 21.8. Notice that these vectors had to be chosen carefully. Given an arbitrary vector, chances are it is not an eigenvector. For example

$$\mathbf{v} = \begin{bmatrix} 1 \\ 1 \end{bmatrix} \qquad \Longrightarrow \qquad A\mathbf{v} = \begin{bmatrix} 1 \\ 0 \end{bmatrix}$$

so $\mathbf{v}$ is not an eigenvector. $\diamond$

This brings up the question of how to find eigenvectors and eigenvalues. We begin with the eigenvalues.

Finding Eigenvalues

Suppose λ is an eigenvalue of a matrix A. Then for some nonzero vector $\mathbf{v}$,

$$A\mathbf{v} = \lambda\mathbf{v}.$$

Putting both terms on the same side we get

$$\lambda\mathbf{v} - A\mathbf{v} = \mathbf{0}.$$

Since $\mathbf{v} = I_n\mathbf{v}$ this becomes

$$(\lambda I_n - A)\mathbf{v} = \mathbf{0}.$$

Since $\mathbf{v}$ is nonzero, this means that the matrix $\lambda I_n - A$ has a nontrivial null space and is therefore not invertible. Thus

$$\det(\lambda I_n - A) = 0.$$

Observe that each step in this chain of implications can be reversed, so we have the following test for eigenvalues.

Proposition 23.1. Let A be an $n \times n$ matrix. Then λ is an eigenvalue of A if and only if $\det(\lambda I_n - A) = 0$.

Example 23.2. Let

$$A = \begin{bmatrix} 1 & 2 \\ 4 & 3 \end{bmatrix}.$$

Then

$$\lambda I_2 - A = \begin{bmatrix} \lambda & 0 \\ 0 & \lambda \end{bmatrix} - \begin{bmatrix} 1 & 2 \\ 4 & 3 \end{bmatrix} = \begin{bmatrix} \lambda - 1 & -2 \\ -4 & \lambda - 3 \end{bmatrix}$$

so

$$\det(\lambda I_2 - A) = (\lambda - 1)(\lambda - 3) - 8 = \lambda^2 - 4\lambda - 5 = (\lambda - 5)(\lambda + 1).$$

This expression is zero when $\lambda = 5$ or $\lambda = -1$, so these are the only eigenvalues of A. $\diamond$

The expression

$$p(\lambda) = \det(\lambda I_n - A)$$

174

is in general a polynomial of degree n, called the **characteristic polynomial** of A. The eigenvalues of A are the roots of this polynomial. If the characteristic polynomial has no real roots then A has no real eigenvalues or eigenvectors.

Example 23.3. The characteristic polynomial of

$$A = \begin{bmatrix} 0 & -1 \\ 1 & 0 \end{bmatrix}$$

is $\lambda^2 + 1$, which has no real roots, so A has no real eigenvalues or eigenvectors. This makes sense since A is the matrix for a rotation (with angle $\pi/2$) which does not send any vectors to multiples of themselves.

$\diamond$

The matrix A does have *complex* eigenvalues. The roots of $\lambda^2 + 1 = 0$ are $\lambda = \pm i$, where $i = \sqrt{-1}$ is an **imaginary number**. A **complex number** is a number of the form $a + bi$ where a and b are real numbers. We will discuss complex eigenvalues in the next section. Next we consider the task of finding the eigenvectors.

Finding Eigenvectors

Once an eigenvalue λ of a matrix A is known, the associated eigenvectors are the nonzero solutions of $A\mathbf{v} = \lambda\mathbf{v}$. Equivalently, they are the nonzero elements of the null space of $\lambda I_n - A$. We call

$$\boxed{E_\lambda = N(\lambda I_n - A)}$$

the **eigenspace** associated with the eigenvalue λ.

Example 23.4. Let $\mathbf{T} : \mathbf{R}^2 \to \mathbf{R}^2$ be the linear transformation with matrix

$$A = \begin{bmatrix} 1 & 2 \\ 4 & 3 \end{bmatrix}.$$

In Example 23.2 we found that the eigenvalues of A were 5 and -1. For $\lambda = 5$,

$$5I_2 - A = \begin{bmatrix} 4 & -2 \\ -4 & 2 \end{bmatrix} \qquad \Longrightarrow \qquad E_5 = N(5I_2 - A) = \text{span}\left(\begin{bmatrix} 1 \\ 2 \end{bmatrix}\right).$$

For $\lambda = -1$,

$$-1I_2 - A = \begin{bmatrix} -2 & -2 \\ -4 & -4 \end{bmatrix} \qquad \Longrightarrow \qquad E_{-1} = N(-1I_2 - A) = \text{span}\left(\begin{bmatrix} -1 \\ 1 \end{bmatrix}\right).$$

Now if we let

$$\mathbf{v}_1 = \begin{bmatrix} 1 \\ 2 \end{bmatrix} \qquad \mathbf{v}_2 = \begin{bmatrix} -1 \\ 1 \end{bmatrix}$$

then $\mathcal{B} = \{\mathbf{v}_1, \mathbf{v}_2\}$ is a basis for $\mathbf{R}^2$ and the matrix for $\mathbf{T}$ with respect to $\mathcal{B}$ is

$$B = C^{-1}AC = \frac{1}{3}\begin{bmatrix} 1 & 1 \\ -2 & 1 \end{bmatrix}\begin{bmatrix} 1 & 2 \\ 4 & 3 \end{bmatrix}\begin{bmatrix} 1 & -1 \\ 2 & 1 \end{bmatrix} = \begin{bmatrix} 5 & 0 \\ 0 & -1 \end{bmatrix}.$$

Notice that the eigenvalues appear on the diagonal.

$\diamond$

Example 23.5. The characteristic polynomial of

$$A = \begin{bmatrix} -1 & 2 & 2 \\ 2 & 2 & -1 \\ 2 & -1 & 2 \end{bmatrix}$$

is

$$p(\lambda) = (\lambda - 3)^2(\lambda + 3),$$

so the eigenvalues are $\lambda = 3$ and $\lambda = -3$. Since

$$3I_3 - A = \begin{bmatrix} 4 & -2 & -2 \\ -2 & 1 & 1 \\ -2 & 1 & 1 \end{bmatrix}$$

the eigenspace associated with $\lambda = 3$ is

$$E_3 = N(3I_3 - A) = \text{span}\left(\begin{bmatrix} 1 \\ 2 \\ 0 \end{bmatrix}, \begin{bmatrix} 1 \\ 0 \\ 2 \end{bmatrix} \right),$$

and since

$$-3I_3 - A = \begin{bmatrix} -2 & -2 & -2 \\ -2 & -5 & 1 \\ -2 & 1 & -5 \end{bmatrix} \implies \text{rref}(-3I_3 - A) = \begin{bmatrix} 1 & 0 & 2 \\ 0 & 1 & -1 \\ 0 & 0 & 0 \end{bmatrix},$$

the eigenspace associated with $\lambda = -3$ is

$$E_{-3} = N(-3I_3 - A) = \text{span}\left(\begin{bmatrix} -2 \\ 1 \\ 1 \end{bmatrix} \right).$$

Letting

$$C = \begin{bmatrix} 1 & 1 & -2 \\ 2 & 0 & 1 \\ 0 & 2 & 1 \end{bmatrix}$$

be the matrix consisting of the basis vectors for the two eigenspaces, we have

$$C^{-1}AC = \begin{bmatrix} 3 & 0 & 0 \\ 0 & 3 & 0 \\ 0 & 0 & -3 \end{bmatrix}.$$

$\diamond$

Diagonalizability

A matrix A is called **diagonalizable** if there exists a matrix C such that $C^{-1}AC$ is a diagonal matrix. That is, A is diagonalizable if A is similar to a diagonal matrix. The matrices in Examples 23.4 and 23.4 are diagonalizable because in both cases consisting of eigenvectors of A.

Proposition 23.2. A real $n \times n$ matrix A is diagonalizable over $\mathbf{R}$ if and only if there exists a basis of $\mathbf{R}^n$ consisting of eigenvectors of A. Such a basis is called an **eigenbasis**.

Proof. First suppose A is diagonalizable, so that $C^{-1}AC = D$ for some diagonal matrix. Since C is invertible, its columns form a basis for $\mathbf{R}^n$. We claim that these columns are all eigenvectors of A. Multiplying both sides by C gives $AC = CD$. Now write

$$C = \begin{bmatrix} | & | & & | \\ \mathbf{v}_1 & \mathbf{v}_2 & \cdots & \mathbf{v}_n \\ | & | & & | \end{bmatrix} \quad \text{and} \quad D = \begin{bmatrix} \lambda_1 & 0 & \cdots & 0 \\ 0 & \lambda_2 & \cdots & 0 \\ \vdots & \vdots & \ddots & \vdots \\ 0 & 0 & \cdots & \lambda_n \end{bmatrix}.$$

Recall that the columns of a product AB are A times the columns of B, so

$$AC = \begin{bmatrix} | & | & & | \\ A\mathbf{v}_1 & A\mathbf{v}_2 & \cdots & A\mathbf{v}_n \\ | & | & & | \end{bmatrix}$$

and

$$CD = \begin{bmatrix} | & | & & | \\ \lambda_1\mathbf{v}_1 & \lambda_2\mathbf{v}_2 & \cdots & \lambda_n\mathbf{v}_n \\ | & | & & | \end{bmatrix}$$

and thus $A\mathbf{v}_i = \lambda_i \mathbf{v}_i$ for $1 \leq i \leq n$.

Now suppose that $\{\mathbf{v}_1, \dots, \mathbf{v}_n\}$ is a basis for $\mathbf{R}^n$ consisting of eigenvectors for A, and let C be the matrix whose columns are these eigenvectors. Then, by the same calculations, $AC = CD$, where D is the diagonal matrix whose diagonal entries are the associated eigenvalues. Since the columns of C form a basis for $\mathbf{R}^n$, C is invertible, and we can solve to get $C^{-1}AC = D$, so A is diagonalizable. $\square$

Example 23.6. The characteristic polynomial of

$$A = \begin{bmatrix} 3 & 0 \\ 2 & 3 \end{bmatrix}$$

is $p(\lambda) = (\lambda - 3)^2$, so $\lambda = 3$ is the only eigenvalue. Since

$$\mathrm{rref}(3I_2 - A) = \begin{bmatrix} 1 & 0 \\ 0 & 0 \end{bmatrix}$$

we have

$$E_3 = \text{span}\left(\begin{bmatrix} 0 \\ 1 \end{bmatrix}\right).$$

Since A has no other eigenvectors, A does not have an eigenbasis, and is therefore not diagonalizable. $\diamond$

The problem in the previous example was caused by the fact that $\lambda = 3$ was a repeated root of the characteristic polynomial. In the case that we do not have repeated roots, the eigenvalues are all different, and it turns out that this implies diagonalizability.

Proposition 23.3. Let A be an $n \times n$ matrix with n different eigenvalues $\lambda_1, \ldots, \lambda_n$. Then A is diagonalizable.

Proof. Choose eigenvector $\mathbf{v}_1, \ldots, \mathbf{v}_n$ with eigenvalues $\lambda_1 \ldots, \lambda_n$, respectively. We claim that these eigenvectors are linearly independent. Suppose

$$c_1\mathbf{v}_1 + c_2\mathbf{v}_2 + \cdots + c_n\mathbf{v}_n = \mathbf{0}.$$

Since $\mathbf{v}_i$ is in the null space of $\lambda_i I_n - A$, we have $(\lambda_i I_n - A)\mathbf{v}_i = \mathbf{0}$. On the other hand $(\lambda_i I_n - A)\mathbf{v}_j = (\lambda_i - \lambda_j)\mathbf{v}_j$ for $j \neq i$. So if we apply the product

$$(\lambda_2 I_n - A) \cdot (\lambda_3 I_n - A) \cdots (\lambda_n I_n - A)$$

to both sides, every term but the $\mathbf{v}_1$ term vanishes, and we are left with

$$(\lambda_2 - \lambda_1)(\lambda_3 - \lambda_1) \cdots (\lambda_n - \lambda_1)c_1\mathbf{v}_1 = \mathbf{0}.$$

Since $\mathbf{v}_1$ is an eigenvector, it is nonzero, and by assumption the eigenvalues are different, so each term in parentheses is nonzero. Hence c_1 must equal zero. Now we are left with

$$c_2\mathbf{v}_2 + \cdots + c_n\mathbf{v}_n = \mathbf{0},$$

and proceeding in the same way we find that $c_2 = 0$, and so on, until $c_n = 0$, so the eigenvectors are linearly independent. Since there are n of them, they form a basis for $\mathbf{R}^n$. $\square$

Example 23.7. The characteristic polynomial of

$$A = \begin{bmatrix} 3 & 8 & 3 \\ 0 & -4 & 5 \\ 0 & 0 & 7 \end{bmatrix}$$

is $p(\lambda) = (\lambda - 3)(\lambda + 4)(\lambda - 7)$, so the eigenvalues are 3, -4 and 7. Since these are different, Proposition 23.3 implies that A is diagonalizable. $\diamond$

Note that Proposition 23.3 provides a sufficient condition for diagonalizability. It is certainly not necessary that the eigenvalues be different. For instance, consider any scalar multiple of the identity, $A = \alpha I_n$. Its characteristic polynomial is $(\lambda - \alpha)^n$, so α is the only eigenvalue, but A is a diagonal matrix, and hence trivially diagonalizable. In Section 25 we will consider another condition which guarantees diagonalizability.

Exercises

23.1. Show that a square matrix A is invertible if and only if zero is not an eigenvalue of A.

23.2. Let A be an invertible matrix, and suppose that $\mathbf{v}$ is an eigenvector of A with eigenvalue λ. Show that $\mathbf{v}$ is an eigenvector of A^{-1}. What is the associated eigenvalue?

23.3. Let V be a subspace of $\mathbf{R}^n$, and let A be a matrix whose columns form a basis for V. Recall that $P = A(A^T A)^{-1} A^T$ is the matrix for the orthogonal projection onto V.

   (a) Show that $P^2 = P$.

   (b) Show that the only eigenvalues of P are 0 and 1.

   (c) What are the associated eigenspaces?

For each matrix A in Exercises 4 through 9, find the eigenvalues and a basis for each corresponding eigenspaces.

23.4. $A = \begin{bmatrix} 1 & 2 \\ 3 & 4 \end{bmatrix}$

23.5. $A = \begin{bmatrix} 1 & 1 \\ 1 & 1 \end{bmatrix}$

23.6. $A = \begin{bmatrix} 0 & 4 \\ 9 & 0 \end{bmatrix}$

23.7. $A = \begin{bmatrix} 2 & 1 & -1 \\ 0 & -1 & 3 \\ 0 & 0 & 3 \end{bmatrix}$

23.8. $A = \begin{bmatrix} 1 & -1 & -1 \\ -1 & 1 & -1 \\ -1 & -1 & 1 \end{bmatrix}$

23.9. $A = \begin{bmatrix} 1 & 1 \\ 1 & 0 \end{bmatrix}$

23.10. Suppose A and B are similar matrices. That is, $B = C^{-1}AC$ for some matrix C.

   (a) Show that A and B have the same characteristic polynomial, and therefore the same eigenvectors.

(b) Show that if $\mathbf{v}$ is an eigenvector of B with eigenvalue λ, then $C\mathbf{v}$ is an eigenvector of A with eigenvalue λ.

23.11. Show that the product of two diagonalizable matrices is diagonalizable.

23.12. Find two diagonalizable matrices whose sum is not diagonalizable.

24 Complex Eigenvalues and Eigenvectors

Recall that the eigenvalues of a matrix A are the roots of its characteristic polynomial $p(\lambda) = \det(\lambda I_n - A)$. As we have seen, not every polynomial has real roots. The polynomial $p(\lambda) = \lambda^2 + 1$, for instance, has roots $\pm i$, where $i = \sqrt{-1}$. Any number of the form bi for some nonzero real number b is called an **imaginary number**. A **complex number** is any number of the form $a + bi$ where a and b are real numbers. We denote by $\mathbf{C}$ the set of all complex numbers. That is,

$$\mathbf{C} = \{a + bi \mid a, b \in \mathbf{R}\}.$$

The complex number $z = a + bi$ is said to have **real part** $\Re(z) = a$ and **imaginary part** $\Im(z) = b$. Thus a real number is simply a complex number with zero imaginary part. A complex number $z = a + bi$ is usually represented by the point (a, b) in the **complex plane**, where the horizontal axis represents the real part and the vertical axis the imaginary part.

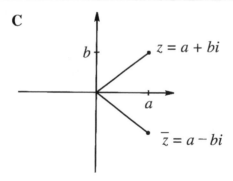

The **complex conjugate** of a complex number $z = a + bi$ is the number

$$\bar{z} = a - bi$$

which, in the complex plane, is the reflection of z across the real axis. Conjugating twice thus does nothing: $\bar{\bar{z}} = z$. A number z is real if and only if $\bar{z} = z$.

Example 24.1. $\overline{2 - 3i} = 2 + 3i$, $\overline{4i} = -4i$, and $\overline{7} = 7$. $\diamond$

The sum of two complex numbers $z_1 = a + bi$ and $z_2 = c + di$ are given by

$$z_1 + z_2 = (a + bi) + (c + di) = (a + b) + (c + d)i$$

and their product is

$$z_1 z_2 = (a + bi)(c + di) = (ac - bd) + (ad + bc)i$$

where in the second formula the fact that $i^2 = -1$ is used.

Example 24.2. $(2 + 3i) + (3 - 4i) = 5 - i$, $(2 + 3i)(3 + 2i) = 13i$, and $(2 + 3i)^2 = -5 + 12i$.
◇

It is easy to check that, for any two complex numbers z_1 and z_2,

$$\overline{z_1 + z_2} = \overline{z_1} + \overline{z_2}$$

and

$$\overline{z_1 z_2} = \overline{z_1}\ \overline{z_2}.$$

The **modulus** of a complex number $z = a + bi$ is $|z| = \sqrt{a^2 + b^2}$, the distance from z to 0 in the complex plane. Notice that the modulus squared can be expressed as

$$|z|^2 = z\bar{z},$$

the product of z and its complex conjugate $\bar{z}$. It is also fairly easy to see that

$$|z_1 z_2| = |z_1||z_2|$$

for any two complex numbers z_1 and z_2.

The formula

$$e^{i\theta} = \cos\theta + i\sin\theta$$

is often useful in simplifying expressions involving complex numbers.

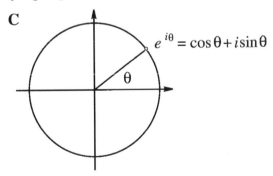

Formally, this comes from the Taylor series expansions of e^x, $\cos x$ and $\sin x$. See the exercises. Any complex number $z = a + bi$ can be expressed as $z = re^{i\theta} = r(\cos\theta + i\sin\theta)$ for some positive real number r, and some angle θ. Taking the modulus of both sides gives

$$r = \sqrt{a^2 + b^2} = |z|$$

The angle θ is determined by looking at a and b. For instance, if $a \neq 0$, then $b/a = (r\sin\theta)/(r\cos\theta) = \tan\theta$.

Example 24.3. Let $z = 1 + i$. Then $|z| = \sqrt{2}$, and $\tan\theta = 1$, so $\theta = \pi/4$. Thus $z = \sqrt{2}(\cos(\pi/4) + i\sin(\pi/4)) = \sqrt{2}e^{i\pi/4}$. Suppose we want to know z^{17}. Then

$$z^{17} = (\sqrt{2})^{17}e^{17i\pi/4}$$
$$= 2^8\sqrt{2}(\cos(17\pi/4) + i\sin(17\pi/4))$$
$$= 256\sqrt{2}(\cos(\pi/4) + i\sin(\pi/4))$$
$$= 256\sqrt{2}(\sqrt{2}/2 + i\sqrt{2}/2)$$
$$= 256(1 + i).$$

$\diamond$

For our purposes the most important fact regarding complex numbers is that any polynomial with complex coefficients factors completely. The proof of this fact requires techniques from complex analysis, and is beyond the scope of this text.

Proposition 24.1. (Fundamental Theorem of Algebra) Let

$$p(z) = a_n z^n + a_{n-1}z^{n-1} + \cdots + a_1 z + a_0$$

be any n^{th} degree polynomial with complex coefficients a_0 through a_n ($a_n \neq 0$). Then

$$p_n(z) = a_n(z - r_1)(z - r_2)\cdots(z - r_n)$$

for some complex numbers r_1 through r_n.

Thus any polynomial with complex coefficients has n complex roots r_1 through r_n. The roots may or may not be distinct. The number of times a root appears in the factorization is called the **multiplicity** of the root. We remark that, since real numbers are complex numbers (with zero imaginary part), this result applies to polynomials with real coefficients as well.

Example 24.4. By the quadratic formula, the polynomial $p(z) = z^2 - 2z + 2$ has roots $1 + i$ and $1 - i$, so $p(z) = (z - 1 - i)(z - 1 + i)$. $\diamond$

Example 24.5. By inspection, the polynomial $p(z) = z^4 - 2z^3 + 2z^2 - 2z + 1$ has a root $z = 1$, so we can factor out $z - 1$ to get $p(z) = (z - 1)(z^3 - z^2 + z - 1)$. Again, by inspection, $z = 1$ is a root of the cubic term, so $p(z) = (z - 1)^2(z^2 + 1)$. Finally, the roots of $z^2 + 1$ are $\pm i$, so $p(z) = (z - 1)^2(z - i)(z + i)$. Thus the roots of $p(z)$ are $z = 1$, with multiplicity 2, $z = i$ and $z = -i$, each with multiplicity 1. $\diamond$

Example 24.6. Let $p(z) = z^2 + i$. The roots are the square roots of $-i$. The number $-i$ lies on the unit circle in the complex plane, at angle $3\pi/2$, so $-i = e^{3\pi i/2}$. Note that if $z = re^{i\theta}$, then $z^2 = r^2 e^{2i\theta}$, so if we choose $r = 1$ and $\theta = 3\pi/4$, then $z^2 = -i$. Thus

$$z_1 = e^{3\pi i/4} = \cos(3\pi/4) + i\sin(3\pi/4) = -\frac{\sqrt{2}}{2} + \frac{\sqrt{2}}{2}i$$

is one root. Clearly $-z_1$ must be the other, so $p(z) = \left(z + \frac{\sqrt{2}}{2} - \frac{\sqrt{2}}{2}i\right)\left(z - \frac{\sqrt{2}}{2} + \frac{\sqrt{2}}{2}i\right)$. $\diamond$

In polynomials with real coefficients the non-real roots appear in complex conjugate pairs.

Proposition 24.2. Let $p(z) = a_n z^n + a_{n-1} z^{n-1} + \cdots + a_1 z + a_0$ be a polynomial with *real* coefficients a_0 through a_n, and suppose z_0 is any root of p. Then $\bar{z}_0$ is a root of p.

Proof. Since z_0 is a root,

$$p(z_0) = a_n z_0^n + a_{n-1} z_0^{n-1} + \cdots + a_1 z_0 + a_0 = 0.$$

Taking the conjugate of both sides, and using the fact that conjugation commutes with addition and multiplication, and the fact that all of the coefficients are real, we have

$$\begin{aligned}
0 &= \overline{a_n z_0^n + a_{n-1} z_0^{n-1} + \cdots + a_1 z_0 + a_0} \\
&= \overline{a_n z_0^n} + \overline{a_{n-1} z_0^{n-1}} + \cdots + \overline{a_1 z_0} + \overline{a_0} \\
&= a_n \bar{z}_0^n + a_{n-1} \bar{z}_0^{n-1} + \cdots + a_1 \bar{z}_0 + a_0 \\
&= p(\bar{z}_0)
\end{aligned}$$

and thus $\bar{z}_0$ is a root. $\qquad\square$

Note that this statement is of interest only in the case that z_0 is not a real number, since if z_0 is real, $\bar{z}_0 = z_0$.

Complex Vectors and Matrices

By analogy with our definition of $\mathbf{R}^n$, we define $\mathbf{C}^n$ to be the set of ordered n-tuples of complex numbers. Vectors in $\mathbf{C}^n$ take the form

$$\mathbf{v} = \begin{bmatrix} v_1 \\ v_2 \\ \vdots \\ v_n \end{bmatrix} = \begin{bmatrix} a_1 + b_1 i \\ a_2 + b_2 i \\ \vdots \\ a_n + b_n i \end{bmatrix}.$$

Addition and scalar multiplication are defined componentwise as in $\mathbf{R}^n$, but now the scalars may be complex numbers as well.

Example 24.7. The vectors

$$\mathbf{v} = \begin{bmatrix} 2 - i \\ -1 + 2i \\ 3 + 3i \end{bmatrix} \quad \text{and} \quad \mathbf{w} = \begin{bmatrix} -1 + 3i \\ 4 + 5i \\ 2 - 5i \end{bmatrix}$$

are elements of $\mathbf{C}^3$. Their sum is

$$\mathbf{v} + \mathbf{w} = \begin{bmatrix} 1 + 2i \\ 3 + 7i \\ 5 - 2i \end{bmatrix}$$

183

and if we let $c = i + 1$, then

$$c\mathbf{v} = \begin{bmatrix} 3 + i \\ -3 + i \\ 6i \end{bmatrix}.$$

◇

With these definitions, the notions of linear combinations, spans, linear independence, subspaces, and linear transformations are defined in the same manner as in $\mathbf{R}^n$. The matrix for a linear transformation $\mathbf{T} : \mathbf{C}^n \to \mathbf{C}^m$ is an $m \times n$ matrix with complex entries. The operations of addition and multiplication of matrices are defined as for real matrices.

Example 24.8. Let

$$A = \begin{bmatrix} 3 + i & 5i \\ -2 & 2 - 2i \end{bmatrix}, \qquad B = \begin{bmatrix} 3 & 1 - 3i \\ 4 + i & 0 \end{bmatrix}, \qquad \text{and} \qquad \mathbf{v} = \begin{bmatrix} 3 + 4i \\ 1 - i \end{bmatrix}.$$

Then

$$A + B = \begin{bmatrix} 6 + i & 1 + 2i \\ 2 + i & 2 - 2i \end{bmatrix}, \qquad AB = \begin{bmatrix} 4 + 23i & 6 - 8i \\ 4 - 6i & -2 + 6i \end{bmatrix}, \qquad \text{and} \qquad A\mathbf{v} = \begin{bmatrix} 10 + 20i \\ -6 - 12i \end{bmatrix}.$$

◇

We define the complex conjugate of a vector in $\mathbf{C}^n$ componentwise by

$$\overline{\mathbf{v}} = \begin{bmatrix} \overline{v_1} \\ \overline{v_2} \\ \vdots \\ \overline{v_n} \end{bmatrix} = \begin{bmatrix} a_1 - b_1 i \\ a_2 - b_2 i \\ \vdots \\ a_n - b_n i \end{bmatrix}.$$

Likewise the conjugate of a complex matrix is defined by taking the conjugate of each entry.

$$\overline{\begin{bmatrix} a_{11} & a_{12} & \cdots & a_{1n} \\ a_{21} & a_{22} & \cdots & a_{2n} \\ \vdots & \vdots & \ddots & \vdots \\ a_{m1} & a_{m2} & \cdots & a_{mn} \end{bmatrix}} = \begin{bmatrix} \overline{a_{11}} & \overline{a_{12}} & \cdots & \overline{a_{1n}} \\ \overline{a_{21}} & \overline{a_{22}} & \cdots & \overline{a_{2n}} \\ \vdots & \vdots & \ddots & \vdots \\ \overline{a_{m1}} & \overline{a_{m2}} & \cdots & \overline{a_{mn}} \end{bmatrix}$$

Example 24.9.

$$\overline{\begin{bmatrix} 2 - i \\ -1 + 2i \\ 3 + 3i \end{bmatrix}} = \begin{bmatrix} 2 + i \\ -1 - 2i \\ 3 - 3i \end{bmatrix} \qquad \text{and} \qquad \overline{\begin{bmatrix} 4 + 7i & 2 - 3i \\ -5 + 2i & 6 + 5i \end{bmatrix}} = \begin{bmatrix} 4 - 7i & 2 + 3i \\ -5 - 2i & 6 - 5i \end{bmatrix}$$

◇

Since the conjugate of a product of complex numbers is the product of the conjugates, and likewise for sums, and since the entries in the product matrix AB are sums of products of entries of A and B, it follows that

$$\overline{AB} = \overline{A}\,\overline{B}.$$

The same property holds for matrix-vector products and dot products, since both are special cases of matrix products. The one major difference between $\mathbf{C}^n$ and $\mathbf{R}^n$ is that the length of a complex vector in $\mathbf{C}^n$ is defined to be

$$\|\mathbf{v}\| = \sqrt{|v_1|^2 + |v_2|^2 + \cdots + |v_n|^2}.$$

where, in place of v_i^2 which could be negative for complex v_i, we have $|v_i|^2$ which is always nonnegative.

Example 24.10. The vector

$$\mathbf{v} = \begin{bmatrix} 5 - 2i \\ 1 + 2i \\ -1 + i \end{bmatrix}$$

has length $\|\mathbf{v}\| = \sqrt{(25 + 4) + (1 + 4) + (1 + 1)} = 6.$ $\qquad\qquad\qquad\qquad \diamondsuit$

With this definition, lengths have the same properties as lengths of real vectors, with one important exception. The relation between dot product and length is

$$\|\mathbf{v}\|^2 = \mathbf{v} \cdot \overline{\mathbf{v}}.$$

Eigenvalues and Eigenvectors

Applying Proposition 24.1 to the characteristic polynomial $p(\lambda) = (\lambda I_n - A)$ of some $n \times n$ matrix A, we see that

$$p(\lambda) = (\lambda - \lambda_1)(\lambda - \lambda_2) \cdots (\lambda - \lambda_n) \qquad\qquad (24.1)$$

for some complex numbers $\lambda_1, \cdots, \lambda_n$, which are the eigenvalues of A. The corresponding eigenspaces are defined in the same way as before, only now they may consist of vectors in $\mathbf{C}^n$

Example 24.11. The characteristic polynomial of

$$A = \begin{bmatrix} 0 & -1 \\ 1 & 0 \end{bmatrix}$$

is $p(\lambda) = \lambda^2 + 1$, so the eigenvalues of A are $\lambda = \pm i$. Since

$$\operatorname{rref}(iI_2 - A) = \begin{bmatrix} 1 & -i \\ 0 & 0 \end{bmatrix},$$

185

the eigenspace associated with $\lambda = i$ is

$$E_i = \text{span}\left(\begin{bmatrix} i \\ 1 \end{bmatrix}\right)$$

and since

$$\text{rref}(-iI_2 - A) = \begin{bmatrix} 1 & i \\ 0 & 0 \end{bmatrix},$$

the eigenspace associated with $\lambda = -i$ is

$$E_{-i} = \text{span}\left(\begin{bmatrix} -i \\ 1 \end{bmatrix}\right).$$

$\diamond$

Notice that the two eigenvectors in this example are conjugates of one another. In general, if A is a matrix with real entries, then the conjugate of an eigenvector is also an eigenvector.

Proposition 24.3. Let A be an $n \times n$ matrix with real entries, and suppose $\mathbf{v}$ is an eigenvector of A with eigenvalue λ. Then $\overline{\mathbf{v}}$ is an eigenvector of A with eigenvalue $\overline{\lambda}$.

Proof. Suppose $\mathbf{v}$ is an eigenvector with eigenvalue λ, so that $A\mathbf{v} = \lambda\mathbf{v}$. Taking the complex conjugate of both sides gives $\overline{A\mathbf{v}} = \overline{\lambda\mathbf{v}} = \overline{\lambda}\,\overline{\mathbf{v}}$. Since A has real entries $\overline{A\mathbf{v}} = \overline{A}\,\overline{\mathbf{v}} = A\,\overline{\mathbf{v}}$. Thus we have $A\,\overline{\mathbf{v}} = \overline{\lambda}\,\overline{\mathbf{v}}$, as claimed. $\square$

We next establish an important relationship between the eigenvalues of a matrix A and the determinant of A. By (24.1), evaluating the characteristic polynomial at $\lambda = 0$ gives

$$p(0) = (-\lambda_1)(-\lambda_2)\cdots(-\lambda_n) = (-1)^n \lambda_1 \cdot \lambda_2 \cdots \lambda_n.$$

On the other hand

$$p(0) = \det(-A) = (-1)^n \det(A),$$

and therefore

$$\det(A) = \lambda_1 \cdot \lambda_2 \cdots \lambda_n.$$

Proposition 24.4. Let A be a square matrix. The determinant of A equals the product of the eigenvalues of A.

Another important quantity associated with a square matrix A is its **trace**, defined by

$$\text{tr}(A) = \sum_{i=1}^{n} a_{ii}.$$

That is, the trace of A is the sum of the diagonal entries of A. In the case that A is a diagonal matrix, the eigenvalues are on the diagonal, so the sum of the eigenvalues equals the trace. This relationship turns out to be true generally.

Proposition 24.5. Let A be a square matrix. The trace of A equals the sum of the eigenvalues of A.

Proof. First we observe that

$$p(\lambda) = (\lambda - \lambda_1)(\lambda - \lambda_2) \cdots (\lambda - \lambda_n)$$
$$= \lambda^n - (\lambda_1 + \lambda_2 + \cdots + \lambda_n)\lambda^{n-1} + \text{ lower degree terms}$$

so the coefficient of λ^{n-1} in $p(\lambda)$ is minus the sum of the eigenvalues. We next find this coefficient in terms of A using the fact that $p(\lambda) = \det(\lambda I_n - A)$. First notice that

$$\lambda I_n - A = \begin{bmatrix} \lambda - a_{11} & -a_{12} & \cdots & -a_{1n} \\ -a_{21} & \lambda - a_{22} & \cdots & -a_{2n} \\ \vdots & \vdots & \ddots & \vdots \\ -a_{n1} & -a_{n2} & \cdots & \lambda - a_{nn} \end{bmatrix}.$$

Expanding the determinant across the first row we claim that the term

$$(\lambda - a_{11}) \begin{vmatrix} \lambda - a_{22} & \cdots & -a_{2n} \\ \vdots & \ddots & \vdots \\ -a_{n2} & \cdots & \lambda - a_{nn} \end{vmatrix}$$

is the only term which contributes to the λ^{n-1} term. To see this, observe that all other terms involve determinants of $(n-1) \times (n-1)$ submatrices obtained by removing the first row and a column other than the first column, and therefore have only $n-2$ entries which contain a λ. Thus these terms contribute only to the terms of degree $n-2$ or less. Now, expanding the remaining determinant across the first row, the same reasoning implies that only the first term will lead to a contribution to λ^{n-1}. Repeating these arguments, we see that

$$\det(\lambda I_n - A) = (\lambda - a_{11})(\lambda - a_{22}) \cdots (\lambda - a_{nn}) + \text{ terms of degree } n-2 \text{ or less.}$$

Thus, expanding the first term, we have

$$p(\lambda) = \lambda^n - (a_{11} + a_{22} + \cdots + a_{nn})\lambda^{n-1} + \text{ terms of degree } n-2 \text{ or less,}$$

so the coefficient of λ^{n-1} in $p(\lambda)$ is minus the trace of A. Comparison with the previous calculation proves the result. $\square$

Exercises

In Exercises 1 through 9, compute the given expression.

24.1. $\begin{bmatrix} 4 - 5i \\ 7 + 8i \end{bmatrix} + \begin{bmatrix} 7i \\ 1 + 6i \end{bmatrix}$

24.2. $i \begin{bmatrix} 1 + 2i \\ 3 + 4i \end{bmatrix}$

24.3. $\begin{bmatrix} 2 + 5i \\ 4 + 2i \\ -1 + 8i \end{bmatrix} + \begin{bmatrix} -2 + i \\ 4i \\ 1 - 7i \end{bmatrix}$

24.4. $(i + 3) \begin{bmatrix} 3 + i \\ 2 - 3i \\ -i \end{bmatrix}$

24.5. $\begin{bmatrix} 2 & 1 + i \\ i - 1 & i \\ 0 & 3 \end{bmatrix} \begin{bmatrix} -1 + 2i \\ 3i \end{bmatrix}$

24.6. $\begin{bmatrix} 2 & 1 + i & 0 \\ 1 + i & i + 4 & 1 + 3i \end{bmatrix} \begin{bmatrix} -1 + 2i \\ 3i \\ 1 \end{bmatrix}$

24.7. $\begin{bmatrix} 2 & 1 + i \\ i - 1 & i \\ 0 & 3 \end{bmatrix} \begin{bmatrix} 2 & 1 + i & 0 \\ 1 + i & i + 4 & 1 + 3i \end{bmatrix}$

24.8. $\begin{bmatrix} 2 & 1 + i & 0 \\ 1 + i & i + 4 & 1 + 3i \end{bmatrix} \begin{bmatrix} 2 & 1 + i \\ i - 1 & i \\ 0 & 3 \end{bmatrix}$

24.9. $\overline{\begin{bmatrix} 5 - 2i & 3 + i \\ -2 + 4i & -1 - 7i \end{bmatrix}}^T$

24.10. Let

$$\mathbf{v} = \begin{bmatrix} 1 + 5i \\ 5 + 7i \end{bmatrix},$$

and compute the following.

(a) $\mathbf{v} + \overline{\mathbf{v}}$

(b) $\mathbf{v} - \overline{\mathbf{v}}$

(c) $\mathbf{v} \cdot \mathbf{v}$

(d) $\mathbf{v} \cdot \overline{\mathbf{v}}$

(e) $\|\mathbf{v}\|$

24.11. Let

$$\mathbf{v} = \begin{bmatrix} 1 + 2i \\ 1 - 5i \\ 2 + i \end{bmatrix},$$

and compute the following.

(a) $\mathbf{v} + \overline{\mathbf{v}}$

(b) $\mathbf{v} - \overline{\mathbf{v}}$

(c) $\mathbf{v} \cdot \mathbf{v}$

(d) $\mathbf{v} \cdot \overline{\mathbf{v}}$

(e) $\|\mathbf{v}\|$

24.12. Consider the following Taylor series expansions.

$$e^x = 1 + x + \frac{x^2}{2!} + \frac{x^3}{3!} + \frac{x^4}{4!} + \cdots$$

$$\cos x = 1 - \frac{x^2}{2!} + \frac{x^4}{4!} - \frac{x^6}{6!} + \cdots$$

$$\sin x = x - \frac{x^3}{3!} + \frac{x^5}{5!} - \frac{x^7}{7!} + \cdots$$

Evaluate the first expansion at $x = i\theta$, and simplify using the other two expansions to deduce the formula $e^{i\theta} = \cos\theta + i\sin\theta$.

24.13. Let z_1 and z_2 be any complex numbers.

(a) Show that $\overline{z_1 + z_2} = \overline{z_1} + \overline{z_2}$.

(b) Show that $\overline{z_1 z_2} = \overline{z_1}\,\overline{z_2}$.

(c) Show that $|z_1 z_2| = |z_1|\,|z_2|$.

Find all roots of the following polynomials, and determine the multiplicity of each root.

24.14. $z^2 + 6$

24.15. $z^2 + 2z + 5$

24.16. $z^3 - z^2 + z - 1$

24.17. $z^4 + 2z^2 + 1$

24.18. $z^4 + z^3 + z^2$

24.19. $z^4 + 4$

24.20. $z^2 - 1 - i$

25 Symmetric Matrices

A square matrix A is called **symmetric** if $A^T = A$.

Example 25.1. Let

$$A = \begin{bmatrix} 1 & 2 & -1 \\ 2 & 3 & 4 \\ -1 & 4 & 7 \end{bmatrix} \quad \text{and} \quad B = \begin{bmatrix} 1 & 2 & 3 \\ 4 & 5 & 2 \\ 6 & 4 & 1 \end{bmatrix}.$$

Then

$$A^T = \begin{bmatrix} 1 & 2 & -1 \\ 2 & 3 & 4 \\ -1 & 4 & 7 \end{bmatrix} = A$$

so A is symmetric, but

$$B^T = \begin{bmatrix} 1 & 4 & 6 \\ 2 & 5 & 4 \\ 3 & 2 & 1 \end{bmatrix} \neq B$$

so B is not symmetric. $\diamond$

It turns out that real symmetric matrices are always diagonalizable. Before we can prove this we need the following result about the eigenvalues of a symmetric matrix.

Proposition 25.1. All the eigenvalues of a real symmetric matrix A are real numbers.

Proof. Let λ be an eigenvalue of A, and let $\mathbf{v}$ be an eigenvector with eigenvalue λ. Then by Proposition 24.3 the vector $\overline{\mathbf{v}}$ is also an eigenvector, with eigenvalue $\overline{\lambda}$. We now evaluate the expression $A\mathbf{v} \cdot \overline{\mathbf{v}}$. First, using the fact that $\mathbf{v}$ is an eigenvector, we get

$$A\mathbf{v} \cdot \overline{\mathbf{v}} = \lambda \mathbf{v} \cdot \overline{\mathbf{v}} = \lambda \|\mathbf{v}\|^2.$$

Next, using Proposition 18.2, the symmetry of A and the fact the $\overline{\mathbf{v}}$ is an eigenvector, we have

$$A\mathbf{v} \cdot \overline{\mathbf{v}} = \mathbf{v} \cdot A^T \overline{\mathbf{v}} = \mathbf{v} \cdot A\overline{\mathbf{v}} = \mathbf{v} \cdot \overline{\lambda}\overline{\mathbf{v}} = \overline{\lambda}\mathbf{v} \cdot \overline{\mathbf{v}} = \overline{\lambda}\|\mathbf{v}\|^2.$$

Since $\mathbf{v}$ is an eigenvector, $\mathbf{v}$ is nonzero, and therefore we have $\lambda = \overline{\lambda}$, so λ is real. $\square$

Proposition 25.2. (Spectral Theorem) Let A be an $n \times n$ real symmetric matrix. Then there exists an orthonormal basis $\mathcal{B} = \{\mathbf{v}_1, \dots, \mathbf{v}_n\}$ of $\mathbf{R}^n$ consisting of eigenvectors of A.

So not only is A diagonalizable, but the eigenvectors can be chosen to be *orthonormal*. If we let C be the matrix whose columns are the basis vectors, then C is an orthogonal matrix, so $C^{-1} = C^T$. It then follows from Proposition 23.2 that

$$C^T A C = D$$

where D is the diagonal matrix whose diagonal entries are the eigenvalues.

Proof. We use induction on the size n of the matrix A. (See Appendix A.) For the case $n = 1$, suppose A is a 1×1 matrix $[\lambda]$, so the vector $[1]$ is a unit eigenvector of A with eigenvalue λ, and forms an orthonormal eigenbasis for $\mathbf{R}^1$. Now suppose the result holds for all $k \times k$ real symmetric matrices. That is, suppose that every $k \times k$ real symmetric matrix has an orthonormal eigenbasis. We must now show that the same holds for any $(k+1) \times (k+1)$ real symmetric matrix. So suppose A is a $(k+1) \times (k+1)$ real symmetric matrix. By Proposition 25.1, the eigenvalues of A are real, so let λ be a real eigenvalue of A, and let $\mathbf{v}$ be a real unit eigenvector with eigenvalue λ. Let $V = \text{span}(\mathbf{v})$ and let $\mathbf{T} : \mathbf{R}^n \to \mathbf{R}^n$ be the linear transformation defined by $\mathbf{T}(\mathbf{x}) = A\mathbf{x}$. We claim that $\mathbf{T}$ sends vectors from $V^\perp$ to $V^\perp$. To see this, suppose $\mathbf{w}$ is in $V^\perp$, so that $\mathbf{w} \cdot \mathbf{v} = 0$. Then, by Proposition 18.2

$$\mathbf{T}(\mathbf{w}) \cdot \mathbf{v} = A\mathbf{w} \cdot \mathbf{v} = \mathbf{w} \cdot A\mathbf{v} = \mathbf{w} \cdot \lambda\mathbf{v} = \lambda\mathbf{w} \cdot \mathbf{v} = 0,$$

so $\mathbf{T}(\mathbf{w})$ is orthogonal to $\mathbf{v}$, and hence orthogonal to any scalar multiple of $\mathbf{v}$. Thus $\mathbf{T}(\mathbf{w})$ is in $V^\perp$. Thus we may regard $\mathbf{T}$ as a linear transformation from $V^\perp$ to $V^\perp$. Since V has dimension 1, $V^\perp$ has dimension k. Let $\mathcal{B} = \{\mathbf{w}_1, \ldots, \mathbf{w}_k\}$ be any orthonormal basis for $V^\perp$ and let B be the matrix for $\mathbf{T} : V^\perp \to V^\perp$ with respect to $\mathcal{B}$. That is, B is the $k \times k$ matrix such that

$$[\mathbf{T}(\mathbf{w})]_\mathcal{B} = B[\mathbf{w}]_\mathcal{B}$$

for all $\mathbf{w}$ in $V^\perp$. Observe that B is a real matrix. We will now show that B is symmetric. Let C denote the matrix whose columns are $\mathbf{w}_1$ through $\mathbf{w}_k$. Since $\mathbf{w} = C[\mathbf{w}]_\mathcal{B}$ for all $\mathbf{w}$ in $V^\perp$, using Proposition 18.2 and the fact that $C^T C = I_k$, we have

$$\mathbf{w} \cdot \mathbf{x} = C[\mathbf{w}]_\mathcal{B} \cdot C[\mathbf{x}]_\mathcal{B} = [\mathbf{w}]_\mathcal{B} \cdot C^T C[\mathbf{x}]_\mathcal{B} = [\mathbf{w}]_\mathcal{B} \cdot [\mathbf{x}]_\mathcal{B}$$

for any $\mathbf{w}$ and $\mathbf{x}$ in $V^\perp$. Using Proposition 18.2 again, along with the symmetry of A, we have

$$\mathbf{T}(\mathbf{w}) \cdot \mathbf{x} = A\mathbf{w} \cdot \mathbf{x} = \mathbf{w} \cdot A\mathbf{x} = \mathbf{w} \cdot \mathbf{T}(\mathbf{x})$$

and thus

$$B[\mathbf{w}]_\mathcal{B} \cdot [\mathbf{x}]_\mathcal{B} = [\mathbf{T}(\mathbf{w})]_\mathcal{B} \cdot [\mathbf{x}]_\mathcal{B} = \mathbf{T}(\mathbf{w}) \cdot \mathbf{x} = \mathbf{w} \cdot \mathbf{T}(\mathbf{x}) = [\mathbf{w}]_\mathcal{B} \cdot [\mathbf{T}(\mathbf{x})]_\mathcal{B} = [\mathbf{w}]_\mathcal{B} \cdot B[\mathbf{x}]_\mathcal{B}$$

for every $\mathbf{w}$ and $\mathbf{x}$ in $V^\perp$. Applying this to the basis vectors in $\mathcal{B}$, we find that

$$B\mathbf{e}_i \cdot \mathbf{e}_j = B[\mathbf{v}_i]_\mathcal{B} \cdot [\mathbf{v}_j]_\mathcal{B} = [\mathbf{v}_i]_\mathcal{B} \cdot B[\mathbf{v}_j]_\mathcal{B} = \mathbf{e}_i \cdot B\mathbf{e}_j$$

for $1 \leq i \leq k$ and $1 \leq j \leq k$. Since $B\mathbf{e}_i$ is column i of B, $B\mathbf{e}_i \cdot \mathbf{e}_j$ is the entry in row j, column i of B. Likewise, $\mathbf{e}_i \cdot B\mathbf{e}_j$ is the entry in row i, column j of B. Since these are equal, B must be symmetric. Since B is $k \times k$, the induction hypothesis implies that there exists an orthonormal basis $\{\mathbf{y}_1, \dots, \mathbf{y}_k\}$ of $\mathbf{R}^k$ consisting of eigenvectors of B. Call the corresponding eigenvectors λ_1 through λ_k and define

$$\mathbf{v}_1 = C\mathbf{y}_1, \qquad \dots \qquad , \mathbf{v}_k = C\mathbf{y}_k.$$

Observing that this implies $[\mathbf{v}_i]_{\mathcal{B}} = \mathbf{y}_i$ for $1 \leq i \leq k$, we have

$$A\mathbf{v}_i = \mathbf{T}(\mathbf{v}_i) = C[\mathbf{T}(\mathbf{v}_i)]_{\mathcal{B}} = CB[\mathbf{v}_i]_{\mathcal{B}} = CB\mathbf{y}_i = C(\lambda_i\mathbf{y}_i) = \lambda_i C\mathbf{y}_i = \lambda_i \mathbf{v}_i$$

so each $\mathbf{v}_i$ is an eigenvector of A. We now claim that $\{\mathbf{v}, \mathbf{v}_1, \dots, \mathbf{v}_k\}$ is an orthonormal basis for $\mathbf{R}^{k+1}$. To see this, we first observe that, since $\{\mathbf{y}_1, \dots, \mathbf{y}_k\}$ is an orthonormal set, the relation

$$\mathbf{v}_i \cdot \mathbf{v}_j = C\mathbf{y}_i \cdot C\mathbf{y}_j = \mathbf{y}_i \cdot C^T C\mathbf{y}_j = \mathbf{y}_i \cdot \mathbf{y}_j$$

implies that $\{\mathbf{v}_1, \dots, \mathbf{v}_k\}$ is an orthonormal set. Next, since the columns of C are in $V^\perp$, each vector $\mathbf{v}_i = C\mathbf{y}_i$ is in $V^\perp$, and therefore orthogonal to $\mathbf{v}$. Hence, since $\mathbf{v}$ is a unit eigenvector of A, $\{\mathbf{v}, \mathbf{v}_1, \dots, \mathbf{v}_k\}$ is an orthonormal basis for $\mathbf{R}^{k+1}$ consisting of eigenvectors of A. By the principle of induction this completes the proof. $\square$

Geometrically, the Spectral Theorem says that any symmetric matrix represents a transformation which stretches or contracts vectors along some set of perpendicular axes.

Example 25.2. Let

$$A = \begin{bmatrix} 1 & 2 \\ 2 & -2 \end{bmatrix}.$$

The characteristic polynomial of A is

$$\lambda^2 + \lambda - 6 = (\lambda + 3)(\lambda - 2)$$

so the eigenvalues of A are $\lambda = -3$ and $\lambda = 2$. Since

$$-3I_2 - A = \begin{bmatrix} -4 & -2 \\ -2 & -1 \end{bmatrix}$$

the eigenspace associated with $\lambda = 4$ is

$$E_{-3} = \text{span}\left(\begin{bmatrix} 1 \\ -2 \end{bmatrix} \right)$$

and since

$$2I_2 - A = \begin{bmatrix} 1 & -2 \\ -2 & 4 \end{bmatrix}$$

the eigenspace associated with $\lambda = 2$ is

$$E_2 = \text{span}\left(\begin{bmatrix} 2 \\ 1 \end{bmatrix}\right).$$

Thus if we let

$$\mathbf{v}_1 = \frac{1}{\sqrt{5}}\begin{bmatrix} 1 \\ -2 \end{bmatrix} \qquad \mathbf{v}_2 = \frac{1}{\sqrt{5}}\begin{bmatrix} 2 \\ 1 \end{bmatrix}$$

then $\mathcal{B} = \{\mathbf{v}_1, \mathbf{v}_2\}$ is an orthonormal eigenbasis for A. The change of basis matrix is

$$C = \frac{1}{\sqrt{5}}\begin{bmatrix} 1 & 2 \\ -2 & 1 \end{bmatrix}$$

and we have

$$C^T A C = \begin{bmatrix} -3 & 0 \\ 0 & 2 \end{bmatrix}.$$

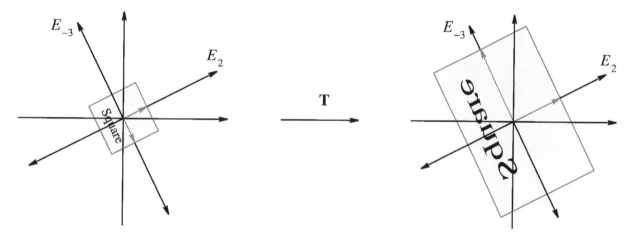

As shown above, the image under the linear transformation defined by A of a square whose sides are parallel to the eigenspaces is a rectangle whose sides are parallel to the eigenspaces. The square is stretched by a factor of 2 parallel to E_2, stretched by a factor of 3 parallel to E_{-3}, and reflected across E_2 (the orthogonal complement of E_{-3}) because of the sign of the eigenvalue -3. $\diamond$

The Spectral Theorem only guarantees the *existence* of an orthonormal basis of eigenvectors. In many cases, however, finding an orthonormal basis can be tricky. The following result simplifies matters somewhat.

Proposition 25.3. Let A be a real symmetric matrix and suppose that $\mathbf{v}_1$ and $\mathbf{v}_2$ are eigenvectors with different eigenvalues $\lambda_1 \neq \lambda_2$. Then $\mathbf{v}_1$ and $\mathbf{v}_2$ are orthogonal.

Proof. Since A is symmetric, the eigenvalues and associated eigenvectors are real. To prove the result we calculate the dot product $A\mathbf{v}_1 \cdot \mathbf{v}_2$ in two ways. First, since $\mathbf{v}_1$ is an eigenvector,

$$A\mathbf{v}_1 \cdot \mathbf{v}_2 = \lambda_1 \mathbf{v}_1 \cdot \mathbf{v}_2.$$

On the other hand, since $\mathbf{v}_2$ is an eigenvector, and A is symmetric, using Proposition 18.2 gives

$$A\mathbf{v}_1 \cdot \mathbf{v}_2 = \mathbf{v}_1 \cdot A^T\mathbf{v}_2 = \mathbf{v}_1 \cdot A\mathbf{v}_2 = \lambda_2 \mathbf{v}_1 \cdot \mathbf{v}_2.$$

Thus, subtracting the two displayed equations results in

$$(\lambda_1 - \lambda_2)\mathbf{v}_1 \cdot \mathbf{v}_2 = 0,$$

and since $\lambda_1 \neq \lambda_2$ by assumption, $\mathbf{v}_1 \cdot \mathbf{v}_2 = 0$, so $\mathbf{v}_1$ and $\mathbf{v}_2$ are orthogonal. $\square$

So eigenvectors with different eigenvalues are automatically orthogonal. The difficulty arises when one or more of the eigenspaces has dimension greater than one. Eigenvectors in the same eigenspace are not necessarily orthogonal. However, by applying the Gram-Schmidt process it is always possible to find an orthonormal basis for each eigenspace.

Example 25.3. The characteristic polynomial of the symmetric matrix

$$A = \begin{bmatrix} 2 & 1 & 1 \\ 1 & 2 & 1 \\ 1 & 1 & 2 \end{bmatrix}$$

is $p(\lambda) = (\lambda - 1)^2(\lambda - 4)$, so the eigenvalues are $\lambda = 1$, with multiplicity 2, and $\lambda = 4$. Since

$$4I_3 - A = \begin{bmatrix} 2 & -1 & -1 \\ -1 & 2 & -1 \\ -1 & -1 & 2 \end{bmatrix} \implies \text{rref}(I_3 - A) = \begin{bmatrix} 1 & 0 & -1 \\ 0 & 1 & -1 \\ 0 & 0 & 0 \end{bmatrix},$$

the vector

$$\mathbf{u}_1 = \frac{1}{\sqrt{3}} \begin{bmatrix} 1 \\ 1 \\ 1 \end{bmatrix}$$

forms an orthonormal basis for E_4. Next, since

$$1I_3 - A = \begin{bmatrix} -1 & -1 & -1 \\ -1 & -1 & -1 \\ -1 & -1 & -1 \end{bmatrix} \implies \text{rref}(1I_3 - A) = \begin{bmatrix} 1 & 1 & 1 \\ 0 & 0 & 0 \\ 0 & 0 & 0 \end{bmatrix},$$

the vectors

$$\mathbf{v}_2 = \begin{bmatrix} -1 \\ 1 \\ 0 \end{bmatrix} \quad \text{and} \quad \mathbf{v}_3 = \begin{bmatrix} -1 \\ 0 \\ 1 \end{bmatrix}$$

form a basis for E_1. Notice that $\mathbf{v}_1$ and $\mathbf{v}_2$ are both orthogonal to $\mathbf{u}_1$ as guaranteed by Proposition 25.3. However $\mathbf{v}_1$ and $\mathbf{v}_2$ are not orthogonal to one another. In Example 22.6, we found via the Gram-Schmidt process that the vectors

$$\mathbf{u}_2 = \frac{1}{\sqrt{2}}\begin{bmatrix} -1 \\ 1 \\ 0 \end{bmatrix} \qquad \text{and} \qquad \mathbf{u}_3 = \frac{1}{\sqrt{6}}\begin{bmatrix} -1 \\ -1 \\ 2 \end{bmatrix}$$

form a basis for the plane spanned by $\mathbf{v}_1$ and $\mathbf{v}_2$. Therefore $\{\mathbf{u}_2, \mathbf{u}_3\}$ is an orthonormal basis for the eigenspace E_1, and consequently, $\{\mathbf{u}_1, \mathbf{u}_2, \mathbf{u}_3\}$ is an orthonormal eigenbasis for A. $\diamond$

Exercises

25.1. Suppose A and B are $n \times n$ symmetric matrices.

    (a) Show that $A + B$ is symmetric.

    (b) Show that cA is symmetric for any scalar c in $\mathbf{R}$.

    (c) Show by example that AB need not be symmetric.

25.2. Let A be any $m \times n$ matrix. Show that $A^T A$ and AA^T are symmetric.

25.3. Show that if A is an invertible symmetric matrix, then A^{-1} is symmetric.

25.4. A square matrix A is called **skew symmetric** if $A^T = -A$. Show that the only real eigenvalue of a skew symmetric matrix is zero.

25.5. Let A be any square matrix. Show that $A + A^T$ is symmetric.

25.6. Show that for all $n \times n$ symmetric matrices A, the matrix $A^2 + I_n$ is invertible.

Find an orthonormal eigenbasis for each symmetric matrix A in Exercises 7 through 17.

25.7. $A = \begin{bmatrix} 2 & 0 \\ 0 & 5 \end{bmatrix}$

25.8. $A = \begin{bmatrix} 0 & 0 \\ 0 & 0 \end{bmatrix}$

25.9. $A = \begin{bmatrix} 0 & 2 \\ 2 & 0 \end{bmatrix}$

25.10. $A = \begin{bmatrix} 1 & -2 \\ -2 & 4 \end{bmatrix}$

25.11. $A = \begin{bmatrix} 3 & 0 \\ 0 & 3 \end{bmatrix}$

25.12. $A = \begin{bmatrix} 3 & 2 \\ 2 & 1 \end{bmatrix}$

25.13. $A = \begin{bmatrix} 1 & 0 & 0 \\ 0 & 2 & 0 \\ 0 & 0 & 3 \end{bmatrix}$

25.14. $A = \begin{bmatrix} 1 & 0 & 0 \\ 0 & 2 & 0 \\ 0 & 0 & 1 \end{bmatrix}$

25.15. $A = \begin{bmatrix} 0 & 0 & 1 \\ 0 & 2 & 0 \\ 1 & 0 & 0 \end{bmatrix}$

25.16. $A = \begin{bmatrix} 0 & 0 & 0 \\ 0 & 2 & 0 \\ 0 & 0 & 0 \end{bmatrix}$

25.17. $A = \begin{bmatrix} 1 & 0 & 0 & 1 \\ 0 & 0 & 0 & 0 \\ 0 & 0 & 0 & 0 \\ 1 & 0 & 0 & 1 \end{bmatrix}$

26 Quadratic Forms

A **quadratic form** is a function $Q : \mathbf{R}^n \to \mathbf{R}$ given by

$$\boxed{Q(\mathbf{x}) = \mathbf{x}^T A \mathbf{x}}$$

where A is a symmetric matrix.

Example 26.1. Let

$$A = \begin{bmatrix} a & b \\ b & c \end{bmatrix}$$

Then

$$Q(x, y) = \begin{bmatrix} x & y \end{bmatrix} \begin{bmatrix} a & b \\ b & c \end{bmatrix} \begin{bmatrix} x \\ y \end{bmatrix} = \begin{bmatrix} x & y \end{bmatrix} \begin{bmatrix} ax + by \\ bx + cy \end{bmatrix} = ax^2 + 2bxy + cy^2$$

is the quadratic form with matrix A.   ◇

Example 26.2. Let

$$A = \begin{bmatrix} a & b & c \\ b & d & e \\ c & e & f \end{bmatrix}$$

Then

$$Q(x,y,z) = \begin{bmatrix} x & y & z \end{bmatrix} \begin{bmatrix} a & b & c \\ b & d & e \\ c & e & f \end{bmatrix} \begin{bmatrix} x \\ y \\ z \end{bmatrix} = ax^2 + dy^2 + fz^2 + 2bxy + 2cxz + 2eyz$$

is the quadratic form with matrix A. $\diamond$

Every quadratic form is a polynomial consisting only of second degree terms, where the entry a_{ij} in row i column j of A gives rise to the term $a_{ij}x_ix_j$ in the quadratic form. The symmetry of A implies that for $i \neq j$ this is the same as $a_{ji}x_jx_i$, and thus these terms combine to give $2a_{ij}x_ix_j$. Conversely, any polynomial consisting solely of second degree terms is a quadratic form. The ij^{th} entry a_{ij} of the associated matrix A is half the coefficient of x_ix_j for $i \neq j$, while each diagonal entries a_{ii} is the coefficient of x_i^2.

Example 26.3. The polynomial

$$Q(x_1, x_2, x_3, x_4) = x_1^2 - 2x_1x_4 + x_2x_3 + 2x_3^2 - 6x_3x_4 - 7x_4^2$$

is the quadratic form with matrix

$$A = \begin{bmatrix} 1 & 0 & 0 & -1 \\ 0 & 0 & 1/2 & 0 \\ 0 & 1/2 & 2 & -3 \\ -1 & 0 & -3 & -7 \end{bmatrix}.$$

$\diamond$

Definiteness

A quadratic form Q is called **positive definite** if $Q(\mathbf{x}) > 0$ for all $\mathbf{x} \neq \mathbf{0}$, **negative definite** if $Q(\mathbf{x}) < 0$ for all $\mathbf{x} \neq \mathbf{0}$, **positive semidefinite** if $Q(\mathbf{x}) \geq 0$ for all $\mathbf{x}$, **negative semidefinite** if $Q(\mathbf{x}) \leq 0$ for all $\mathbf{x}$, and **indefinite** if there exist $\mathbf{x}_1$ and $\mathbf{x}_2$ such that $Q(\mathbf{x}_1) > 0$ and $Q(\mathbf{x}_2) < 0$.

Example 26.4.

$$A = \begin{bmatrix} 1 & 0 \\ 0 & 2 \end{bmatrix} \qquad \Longrightarrow \qquad Q(x,y) = x^2 + 2y^2$$

It is clear that $Q(x,y) > 0$ for all $(x,y) \neq (0,0)$, so Q is positive definite. The graph of the function Q is shown below.

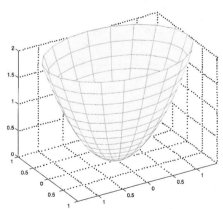

Example 26.5.

$$B = \begin{bmatrix} -1 & 0 \\ 0 & -2 \end{bmatrix} \quad \Longrightarrow \quad Q(x,y) = -x^2 - 2y^2$$

In this case $Q(x,y) < 0$ for all $(x,y) \neq (0,0)$, so Q is negative definite.

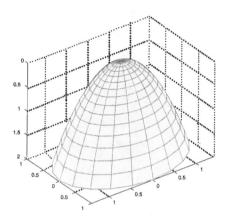

Example 26.6.

$$C = \begin{bmatrix} 1 & 0 \\ 0 & -2 \end{bmatrix} \quad \Longrightarrow \quad Q(x,y) = x^2 - 2y^2$$

Since $Q(1,0) > 0$ and $Q(0,1) < 0$, Q is indefinite.

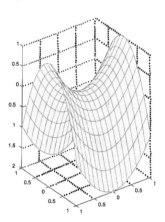

Example 26.7.

$$D = \begin{bmatrix} 1 & 0 \\ 0 & 0 \end{bmatrix} \quad \Longrightarrow \quad Q(x,y) = x^2$$

Since $Q(\mathbf{x}) \geq 0$ for all $\mathbf{x}$, Q is positive semidefinite. Q is not positive definite since $Q(1,0) = 0$.

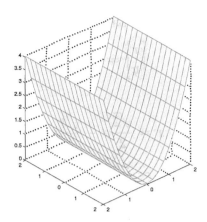

The matrix $-D$ gives rise to a form which is negative semidefinite. $\diamond$

Eigenvalues and Quadratic Forms

There is a beautiful relationship between the eigenvalues of a symmetric matrix and the definiteness of the quadratic form it generates. This is illustrated in the following example.

Example 26.8.

$$A = \begin{bmatrix} 1 & 4 \\ 4 & 1 \end{bmatrix} \qquad \Longrightarrow \qquad Q(x,y) = x^2 + 8xy + y^2$$

At first glance it is not obvious whether this form is positive definite, negative definite, or indefinite. In all the previous examples, the matrices were diagonal matrices, so there was no xy term. By finding the eigenvalues and eigenvectors of A we can make a change of coordinates which will eliminate this cross term. First we find the eigenvalues. Since

$$\det(\lambda I_2 - A) = (\lambda - 1)^2 - 16 = \lambda^2 - 2\lambda - 15 = (\lambda - 5)(\lambda + 3)$$

the eigenvalues are 5 and -3. Next, we find the corresponding eigenspaces.

$$5I_2 - A = \begin{bmatrix} 4 & -4 \\ -4 & 4 \end{bmatrix} \xrightarrow[\text{rref}]{} \begin{bmatrix} 1 & -1 \\ 0 & 0 \end{bmatrix} \qquad \Longrightarrow \qquad E_5 = \text{span}\left(\begin{bmatrix} 1 \\ 1 \end{bmatrix}\right)$$

$$-3I_2 - A = \begin{bmatrix} -4 & -4 \\ -4 & -4 \end{bmatrix} \xrightarrow[\text{rref}]{} \begin{bmatrix} 1 & 1 \\ 0 & 0 \end{bmatrix} \qquad \Longrightarrow \qquad E_{-3} = \text{span}\left(\begin{bmatrix} -1 \\ 1 \end{bmatrix}\right)$$

Now consider a basis consisting of **unit** eigenvectors of A,

$$\left\{ \begin{bmatrix} \frac{\sqrt{2}}{2} \\ \frac{\sqrt{2}}{2} \end{bmatrix}, \begin{bmatrix} -\frac{\sqrt{2}}{2} \\ \frac{\sqrt{2}}{2} \end{bmatrix} \right\}$$

and let

$$C = \begin{bmatrix} \frac{\sqrt{2}}{2} & -\frac{\sqrt{2}}{2} \\ \frac{\sqrt{2}}{2} & \frac{\sqrt{2}}{2} \end{bmatrix}$$

199

denote the change of basis matrix for this basis. If we make the change of variable

$$\begin{bmatrix} x \\ y \end{bmatrix} = C \begin{bmatrix} u \\ v \end{bmatrix}$$

then

$$x = \frac{\sqrt{2}}{2}u - \frac{\sqrt{2}}{2}v$$

$$y = \frac{\sqrt{2}}{2}u + \frac{\sqrt{2}}{2}v.$$

So

$$Q(x,y) = \left(\frac{\sqrt{2}}{2}u - \frac{\sqrt{2}}{2}v \right)^2 + 8\left(\frac{\sqrt{2}}{2}u - \frac{\sqrt{2}}{2}v \right)\left(\frac{\sqrt{2}}{2}u + \frac{\sqrt{2}}{2}v \right) + \left(\frac{\sqrt{2}}{2}u + \frac{\sqrt{2}}{2}v \right)^2$$

$$= 5u^2 - 3v^2.$$

In (u,v) coordinates it is clear that Q is indefinite, since the coefficients have opposite sign. For instance Q is positive when $(u,v) = (1,0)$, and Q is negative when $(u,v) = (0,1)$. Since

$$C(1,0) = (\sqrt{2}/2, \sqrt{2}/2) \qquad \text{and} \qquad C(0,1) = (-\sqrt{2}/2, \sqrt{2}/2)$$

we know the (x,y) coordinates at which Q is positive and negative.

$$Q(\sqrt{2}/2, \sqrt{2}/2) = 5 \qquad Q(-\sqrt{2}/2, \sqrt{2}/2) = -3$$

Notice that the coefficients of u^2 and v^2 are exactly the eigenvalues of A. This is not a coincidence. $\diamond$

Proposition 26.1. Let $Q(\mathbf{x}) = \mathbf{x}^T A \mathbf{x}$ be the quadratic form generated by an $n \times n$ symmetric matrix A.

1. If all of the eigenvalues of A are positive then Q is positive definite.

2. If all of the eigenvalues of A are negative then Q is negative definite.

3. If all of the eigenvalues of A are nonnegative then Q is positive semidefinite.

4. If all of the eigenvalues of A are nonpositive then Q is negative semidefinite.

5. If A has both positive and negative eigenvalues then Q is indefinite.

Proof. By the Spectral Theorem, there exists an orthonormal basis $\{\mathbf{u}_1, \mathbf{u}_2, \ldots, \mathbf{u}_n\}$ of $\mathbf{R}^n$ consisting of eigenvectors of A. Let

$$C = \begin{bmatrix} | & | & & | \\ \mathbf{u}_1 & \mathbf{u}_2 & \cdots & \mathbf{u}_n \\ | & | & & | \end{bmatrix}$$

be the change of basis matrix for this basis. Then we have

$$C^T A C = D = \begin{bmatrix} \lambda_1 & 0 & \cdots & 0 \\ 0 & \lambda_2 & \cdots & 0 \\ \vdots & \vdots & \ddots & \vdots \\ 0 & 0 & \cdots & \lambda_n \end{bmatrix}$$

where $\lambda_1, \lambda_2, \ldots, \lambda_n$ are the eigenvalues of A. If we make the change of variable $\mathbf{x} = C\mathbf{y}$ then

$$\begin{aligned} Q(\mathbf{x}) &= \mathbf{x}^T A \mathbf{x} \\ &= (C\mathbf{y})^T A C \mathbf{y} \\ &= \mathbf{y}^T C^T A C \mathbf{y} \\ &= \mathbf{y}^T D \mathbf{y} \\ &= \lambda_1 y_1^2 + \lambda_2 y_2^2 + \cdots + \lambda_n y_n^2. \end{aligned}$$

It is clear from this expression that $Q(\mathbf{x}) \geq 0$ for all $\mathbf{x}$ if all of the eigenvalues are nonnegative, and that $Q(\mathbf{x}) \leq 0$ for all $\mathbf{x}$ if all of the eigenvalues are nonpositive. This proves statements 3 and 4 in the theorem.

Next suppose that all of the eigenvalues of A are positive. Then $Q(\mathbf{x})$ is positive unless $y_1 = y_2 = \cdots = y_n = 0$, in which case $\mathbf{y} = \mathbf{0}$, which implies $\mathbf{x} = C\mathbf{0} = \mathbf{0}$. Thus Q is positive definite. Similarly, if all of the eigenvalues of A are negative, then Q is negative definite.

Finally, suppose that A has both positive and negative eigenvalues. If $\lambda_i > 0$ and $\lambda_j < 0$, let $\mathbf{y}_1 = \mathbf{e}_i$ and $\mathbf{y}_2 = \mathbf{e}_j$, and let

$$\mathbf{x}_1 = C\mathbf{y}_1 = C\mathbf{e}_i \quad \text{and} \quad \mathbf{x}_2 = C\mathbf{y}_2 = C\mathbf{e}_j.$$

Then

$$Q(\mathbf{x}_1) = \lambda_i > 0 \quad \text{and} \quad Q(\mathbf{x}_2) = \lambda_j < 0$$

so Q is indefinite. $\qquad \square$

Example 26.9. Let

$$A = \begin{bmatrix} 1 & 2 & 2 \\ 2 & 1 & 2 \\ 2 & 2 & 1 \end{bmatrix}.$$

The characteristic polynomial of A is

$$\begin{aligned} p(\lambda) &= (\lambda - 1)[(\lambda - 1)^2 - 4] + 2[(-2)(\lambda - 1) - 4] - 2[4 + 2(\lambda - 1)] \\ &= (\lambda - 1)^3 - 12\lambda - 4 \\ &= \lambda^3 - 3\lambda^2 - 9\lambda - 5 \\ &= (\lambda + 1)^2(\lambda - 5) \end{aligned}$$

The eigenvalues are therefore -1, with multiplicity 2, and 5, so the quadratic form with matrix A is indefinite. $\qquad \diamond$

201

Example 26.10. Let

$$A = \begin{bmatrix} 2 & 2 & 2 \\ 2 & 2 & 2 \\ 2 & 2 & 2 \end{bmatrix}.$$

The characteristic polynomial of A is

$$p(\lambda) = \lambda^3 - 6\lambda^2 = \lambda^2(\lambda - 6)$$

so the eigenvalues are 0, with multiplicity 2, and 6, and the associated quadratic form is positive semidefinite. $\diamond$

Example 26.11. Let

$$A = \begin{bmatrix} 3 & 2 & 2 \\ 2 & 3 & 2 \\ 2 & 2 & 3 \end{bmatrix}.$$

The characteristic polynomial of A is

$$p(\lambda) = (\lambda - 1)^2(\lambda - 7)$$

so the eigenvalues are 1 (with multiplicity 2) and 7, and the associated quadratic form is positive definite. $\diamond$

For a 2×2 symmetric matrix A, we can easily determine the signs of the eigenvalues in terms of the trace and determinant of A.

Proposition 26.2. Let

$$A = \begin{bmatrix} a & b \\ b & c \end{bmatrix}$$

be a 2×2 symmetric matrix, and let Q be the associated quadratic form.

1. If $\det(A) > 0$ and $\operatorname{tr}(A) > 0$ then the eigenvalues of A are both positive, so Q is positive definite.

2. If $\det(A) > 0$ and $\operatorname{tr}(A) < 0$ then the eigenvalues of A are both negative, so Q is negative definite.

3. If $\det(A) = 0$ and $\operatorname{tr}(A) \geq 0$ then the eigenvalues of A are both nonnegative, so Q is positive semidefinite.

4. If $\det(A) = 0$ and $\operatorname{tr}(A) \leq 0$ then the eigenvalues of A are both nonpositive, so Q is negative semidefinite.

5. If $\det(A) < 0$ then the eigenvalues of A have opposite signs, so Q is indefinite.

Proof. If $\det(A) > 0$ then, since $\det(A)$ is the product of the eigenvalues, the eigenvalues are nonzero and have the same sign. Since $\operatorname{tr}(A)$ is the sum of the eigenvalues, this sign is positive if $\operatorname{tr}(A) > 0$ and negative if $\operatorname{tr}(A) < 0$.

If $\det(A) = 0$, at least one of the eigenvalues is zero, and therefore $\operatorname{tr}(A)$ equals the other eigenvalue. So if $\operatorname{tr}(A) \geq 0$ they are both nonnegative, and if $\operatorname{tr}(A) \leq 0$ they are both nonpositive.

Finally, if $\det(M) < 0$ then the product of the eigenvalues is negative, so the eigenvalues have opposite signs. $\qquad\square$

Exercises

In Exercises 1 through 8, find the matrix for the given quadratic form, and determine the definiteness of the form.

26.1. $Q(x, y) = x^2 + 2xy + y^2$

26.2. $Q(x, y, z) = 2xy + 2xz + 2yz$

26.3. $Q(x, y) = 3x^2 - 2xy + 2y^2$

26.4. $Q(x, y, z) = x^2 + y^2$

26.5. $Q(x, y, z) = -2x^2 - 3y^2 - z^2$

26.6. $Q(w, x, y, z) = -2x^2 - 3y^2 - z^2$

26.7. $Q(w, x, y, z) = x^2 + 4yz$

26.8. $Q(w, x, y, z) = 4x^2 + xz$

In Exercises 9 through 18, determine the definiteness of the quadratic form Q with the given matrix A.

26.9. $A = \begin{bmatrix} 5 & 6 \\ 6 & 8 \end{bmatrix}$

26.10. $A = \begin{bmatrix} -1 & -2 \\ -2 & -3 \end{bmatrix}$

26.11. $A = \begin{bmatrix} -1 & -1 \\ -1 & -3 \end{bmatrix}$

26.12. $A = \begin{bmatrix} 1 & -2 \\ -2 & 4 \end{bmatrix}$

26.13. $A = \begin{bmatrix} -1 & 3 \\ 3 & -9 \end{bmatrix}$

26.14. $A = \begin{bmatrix} 2 & -1 \\ -1 & 1 \end{bmatrix}$

26.15. $A = \begin{bmatrix} 1 & 0 & 1 \\ 0 & 1 & 0 \\ 1 & 0 & 1 \end{bmatrix}$

26.16. $A = \begin{bmatrix} 1 & 1 & 1 \\ 1 & 1 & 1 \\ 1 & 1 & 1 \end{bmatrix}$

26.17. $A = \begin{bmatrix} 1 & 0 & 2 \\ 0 & 1 & 0 \\ 2 & 0 & 1 \end{bmatrix}$

26.18. $A = \begin{bmatrix} 1 & -2 & 8 & 3 \\ -2 & 3 & 4 & 7 \\ 8 & 4 & -1 & 2 \\ 3 & 7 & 2 & 5 \end{bmatrix}$ Hint: You do not need to find the eigenvalues.

26.19. Show that if $Q_1 : \mathbf{R}^n \to \mathbf{R}$ and $Q_2 : \mathbf{R}^n \to \mathbf{R}$ are quadratic forms with matrices A_1 and A_2, respectively, then $Q_1 + Q_2$ is a quadratic form with matrix $A_1 + A_2$.

26.20. Let A_1 and A_2 be $n \times n$ symmetric matrices, both of which have only positive eigenvalues. Use the result of the previous exercise to show that all the eigenvalues of $A_1 + A_2$ are positive.

A   Principle of Mathematical Induction

In this section we briefly explain the method of proof by mathematical induction. This method of proof applies to sequences of statements S_n, which are indexed by the positive integers n. The idea of induction is simple. We first prove that S_1 is true, and then that the truth of each statement S_k implies the truth of the next statement S_{k+1}. Having done this, since S_1 is true and S_1 implies S_2, it follows that S_2 is true. Since S_2 is now true and S_2 implies S_3, S_3 must also be true, an so on.

To make this argument somewhat more rigorous we need to use one of the axioms of the positive integers, called the Well Ordering Principle.

> **Well Ordering Principle.**
> Every nonempty set of positive integers contains a smallest element.

The well ordering principle is an axiom, so it does not make sense to speak of proving it. We can however prove the principle of induction using the Well Ordering Principle.

Proposition A.1. (Principle of Mathematical Induction) Let $\{S_n\}_{n=1}^{\infty}$ be a sequence of statements and suppose the following two conditions are satisfied.

1. S_1 is true.

2. For all positive integers k, if S_k is true then S_{k+1} is true.

Then S_n is true for all positive integers n.

Proof. Let $\{S_n\}_{n=1}^{\infty}$ be a sequence of statements which satisfies the two properties above, and let F be the set of positive integers k for which the statement S_k is false. We intend to show that F is the empty set, and therefore S_n is true for all positive integers n. To do so, we suppose to the contrary that F is nonempty. Then by the Well Ordering Principle, F has a smallest element. Call this positive integer m. By property 1, S_1 is true, so 1 is not an element of F, and thus $m > 1$. Since m is the smallest element of F, $m - 1$ is a positive integer which is not in F, and therefore S_{m-1} is true. By property 2, however, this implies that S_m is true, and therefore m is not in F. Thus we have a contradiction, which means that our assumption that F is nonempty is false. $\qquad\square$

Example A.1. Let S_n be the following formula for the sum of the first n positive integers.

$$\sum_{i=1}^{n} i = \frac{n(n+1)}{2}$$

We will use induction to prove this for all positive integers n. First we must establish the base case S_1. Since

$$\sum_{i=1}^{1} i = 1 = \frac{1(1+1)}{2}$$

the statement S_1 is true. Next we must show for all positive integers k that *if S_k is true then S_{k+1} is true*. So we suppose S_k is true. That is we suppose

$$\sum_{i=1}^{k} i = \frac{k(k+1)}{2}$$

for some positive integer k. From this we need to conclude that S_{k+1} is true, i.e. that

$$\sum_{i=1}^{k+1} i = \frac{(k+1)(k+2)}{2}.$$

To do so we notice that

$$\sum_{i=1}^{k+1} i = \sum_{i=1}^{k} i + (k+1),$$

205

so using the hypothesis that S_k is true this becomes

$$\sum_{i=1}^{k+1} i = \frac{k(k+1)}{2} + (k+1) = \frac{1}{2}(k^2 + 3k + 2) = \frac{(k+1)(k+2)}{2},$$

which is what we were trying to show.   ◇

B   Existence of Bases

Proposition B.1. Let V be a nontrivial subspace of $\mathbf{R}^n$. Then there exists a basis for V.

Proof. We first make the following claim. For any set of k linearly independent vectors $\{\mathbf{v}_1, \dots, \mathbf{v}_k\}$ in V, either

1. these vectors form a basis for V, or

2. there exists a vector $\mathbf{v}_{k+1}$ in V such that $\{\mathbf{v}_1, \dots, \mathbf{v}_k, \mathbf{v}_{k+1}\}$ is a linearly independent set.

Before proving the claim, we use it to prove the Proposition. Since V is nontrivial, there exists a nonzero vector $\mathbf{v}_1$ in V. The set $\{\mathbf{v}_1\}$ is linearly independent. If this set is not a basis for V, then by the claim, there exists a vector $\mathbf{v}_2$ in V, such that the set $\{\mathbf{v}_1, \mathbf{v}_2\}$ is linearly independent. Now if this set is not a basis for V, then by the claim, there exists a vector $\mathbf{v}_3$ in V, such that the set $\{\mathbf{v}_1, \mathbf{v}_2, \mathbf{v}_3\}$ is linearly independent. This process must eventually terminate in a basis for V, since otherwise, we could find a set of more than n linearly independent vectors in $\mathbf{R}^n$, and this contradicts Proposition 8.4.

Now we prove the claim. Suppose $\{\mathbf{v}_1, \dots, \mathbf{v}_k\}$ is a linearly independent set, and let $V_k = \text{span}(\mathbf{v}_1, \dots, \mathbf{v}_k)$. If $V_k = V$, then $\{\mathbf{v}_1, \dots, \mathbf{v}_k\}$ is a basis for V. Otherwise, there exists a vector $\mathbf{v}_{k+1}$ in V which is not in V_k. Now suppose

$$c_1\mathbf{v}_1 + \cdots + c_k\mathbf{v}_k + c_{k+1}\mathbf{v}_{k+1} = \mathbf{0}.$$

If $c_{k+1} \neq 0$, then solving for $\mathbf{v}_{k+1}$ gives

$$\mathbf{v}_{k+1} = -\frac{c_1}{c_{k+1}}\mathbf{v}_1 - \cdots - \frac{c_k}{c_{k+1}}\mathbf{v}_k$$

which implies $\mathbf{v}_{k+1}$ is in V_k, a contradiction. Hence c_{k+1} must equal zero. But this implies

$$c_1\mathbf{v}_1 + \cdots + c_k\mathbf{v}_k = \mathbf{0}$$

and the linear independence of $\{\mathbf{v}_1, \dots, \mathbf{v}_k\}$ implies that $c_1 = \cdots = c_k = 0$. Thus the set $\{\mathbf{v}_1, \dots, \mathbf{v}_k, \mathbf{v}_{k+1}\}$ is linearly independent. This proves the claim.   □

C   Uniqueness of Reduced Row Echelon Form

> **Proposition C.1.** Let A be any $m \times n$ matrix. Then A is row equivalent to exactly one reduced row echelon form matrix. This matrix is called the reduced row echelon form of A and is denoted $\mathrm{rref}(A)$.

Proof of Uniqueness. If a matrix A is row equivalent to two reduced row echelon form matrices R_1 and R_2, then R_1 and R_2 are row equivalent to each other. Thus, to prove uniqueness it suffices to show that if two reduced row echelon form matrices are row equivalent, then they must be equal.

We prove this by induction on the number of columns n. For the base case $n = 1$, there are only two possible reduced row echelon form matrices with one column,

$$\begin{bmatrix} 1 \\ 0 \\ \vdots \\ 0 \end{bmatrix} \quad \text{or} \quad \begin{bmatrix} 0 \\ 0 \\ \vdots \\ 0 \end{bmatrix},$$

and it is clear that they are not row equivalent. For our inductive hypothesis, we assume that any two row equivalent, reduced row echelon form matrices with k columns are equal. We now need to show that any two row equivalent, reduced row echelon form matrices R_1 and R_2 with $k + 1$ columns are equal. If we write

$$R_1 = \begin{bmatrix} A & \mathbf{b}_1 \end{bmatrix} \quad \text{and} \quad R_2 = \begin{bmatrix} B & \mathbf{b}_2 \end{bmatrix},$$

where A and B are matrices with k columns and $\mathbf{b}_1$ and $\mathbf{b}_2$ are column vectors, then A and B are in reduced row echelon form, and row equivalent via the same row operations which take R_1 to R_2. So by the induction hypothesis, $A = B$, and we have

$$R_1 = \begin{bmatrix} A & \mathbf{b}_1 \end{bmatrix} \quad \text{and} \quad R_2 = \begin{bmatrix} A & \mathbf{b}_2 \end{bmatrix}.$$

Now these matrices can be regarded as the augmented matrices for the systems $A\mathbf{x} = \mathbf{b}_1$ and $A\mathbf{x} = \mathbf{b}_2$, respectively. Since R_1 and R_2 are row equivalent, the solution sets of these systems are the same. If this common solution set is nonempty, we may consider any such solution $\mathbf{x}_0$. Then $\mathbf{b}_1 = A\mathbf{x}_0 = \mathbf{b}_2$, and therefore $R_1 = R_2$. On the other hand, if neither system has a solution, then by Proposition 6.2, both systems contain the equation $0 = 1$, and therefore have a pivot in the final column. Thus both $\mathbf{b}_1$ and $\mathbf{b}_2$ have a 1 in one component, and a zero in all others. But by definition of reduced row echelon form, the 1 must be in the row immediately below the last pivot of A. This again implies $\mathbf{b}_1 = \mathbf{b}_2$ and thus $R_1 = R_2$. By induction it follows that any two row equivalent reduced row echelon form matrices are equal. $\qquad \square$

Index